# THE LAYMAN'S GUIDE TO CLIMATE CHANGE

## MOHAMMED GHASSAN FARIJA

Cover design by Marko Polic

Edited by Chrissy Cutting & Robin Schroffel

*First Edition*

بسم الله الرحمن الرحيم

*In the name of God, the Most Gracious, the Most Merciful*

*Pollution has appeared in the land and the sea by the hands of mankind for what they earned.*
*- Quran 30:41*

# CONTENTS

# INTRODUCTION

*It is not knowledge which should come to you. It is you who should come to knowledge.*
— *Malik ibn Anas*

Despite the numerous challenges facing humanity today, we are living in what is by most accounts the most prosperous period in human history.

In the early years of the nineteenth century, the vast majority of humanity lived in abject poverty. The technological and economic progress of the last two centuries, sparked by the Industrial Revolution, changed everything. During that time, the world's population increased sevenfold. While many would have thought that might result in even greater poverty, the opposite happened. By the early 1980s, less than half of humanity, 44 percent, lived in abject poverty. A mere three decades later, the figure now sits closer to 10 percent. Never before have so many been lifted from the grips of impoverishment so fast. In fact, the living standards enjoyed by the working and middle classes of today, in many respects, far exceed those experienced by the rich only two centuries ago. To put this in perspective, the wealthiest man in the world in 1836, Nathan Rothschild, died of what would now be considered an easily treatable infection.[1]

Since the end of the Second World War in 1945, the number of deaths caused by violent conflicts has rapidly declined and has reached a historical low.[2] Not only do people enjoy more financially prosperous and peaceful lives than their ancestors did, they also live longer. As industrial development, economic growth and better medicine spread across the globe; life expectancy has more than doubled within the last two centuries. "No country in the world has a lower life expectancy than the countries with the highest life expectancy in 1800."[3] There is, however, a lingering specter which casts its shadow upon all that we have built.

*Climate change.*

It is a complex, all-encompassing, scientific, economic, political and social problem that, for most of us, can seem far removed from our everyday lives. With mortgages, loans, children and our careers, many of us find it hard to put climate change on our list of things to care about, in any practical sense at least. However, we often catch ourselves pondering about the future, trying to imagine the sort of world us and our children will one day live in. Looking at the scientific evidence, it is difficult to imagine a future where anyone's life isn't touched by climate change. It is, without a shadow of a doubt, the existential problem of our age.

It is essential to recognize that climate change is no longer merely a far distant threat. From record-breaking heat waves to increased coastal flooding, the specter has reared its ugly head. While the more extreme consequences are still on the horizon, make no mistake about it, the effects have begun to manifest. The impacts of tomorrow will be decided by the actions of today. If we are to continue on our current trajectory, there is a high probability that we will cross the threshold of what is considered a "safe level" of warming within less than two decades. As the clock ticks away, it is imperative that the general public gains a deep understanding of the issue and what is at stake.

The purpose of this book is to present the issue of climate change in a manner that anyone can understand, without oversimplification or losing the necessary nuance when analyzing the problem. I aim to encompass, between these two covers, the most pertinent information regarding climate change and communicate it in a clear and succinct manner in which a person without a scientific background can easily digest and comprehend. I am confident that some of those reading this book may have some skepticism concerning climate change. To those, I ask for nothing but the ability to set aside any preconceived notions while you are reading this book.

I must admit that in writing this book, providing the public with an easily accessible resource to learn about climate change is not my only goal. Many of those who are, or will be, tasked with determining economic and political policies, rulings, and decisions about climate change do not have a background in science or engineering (nor will many of those who should hold them into account, i.e., the general public). Many of these decision makers will have backgrounds in law, economics or finance. Additionally, as we will come to learn in later sections of this book, the technological solutions needed to mitigate climate change are already well within our grasp. What we lack is an economic and political framework that incentivizes us to put these technologies into practice on a global scale. I hope that this book acts as a foundation of sorts for these individuals from which they might base their decisions.

It is my hope that upon finishing this book, you will come to understand the fundamental mechanisms that govern our climate, whether the recent increase in the globe's average surface temperature is natural or manmade, the causes of climate change in the past, reasons behind why the climate "debate" exists, expected changes and the socioeconomic impacts they entail, the role of political action and climate treaties such as the Paris Climate Accords and the Kyoto Protocol, and perhaps most importantly, the solutions to the problem. We will venture into the distant past as well as many possible futures, from the North Pole to the Sahara Desert. We will cover everything from established science to corporate conspiracy to international politics. The only tools you require for this journey are courage, critical thinking and an open mind.

# 1

## THE THEORY

*It doesn't matter how beautiful your theory is, it doesn't matter how smart you are. If it doesn't agree with experiment, it's wrong.*
    *— Richard P. Feynman*

IT IS safe to say the Industrial Revolution transformed our way of life. Whether it's bottled water, planes, cars, light bulbs, or your smartphone, much of what we take for granted would not exist had it not been for the Industrial Revolution.

Throughout the majority of human history, people lived in the countryside as farmers. Unlike today, most did not sell their produce. They were subsistence farmers who ate what they grew. Most were poor, hunger was prevalent, and diseases spread like wildfire. The Industrial Revolution was a period of rapid development, shifting economies from agricultural to industrial. People poured into cities to work in factories in search of a better life. This influx led to many health problems due to a lack of proper sanitation. In addition, they lacked a proper understanding of how diseases spread. However, the Industrial Revolution also armed humanity with the tools of better technology and medicine. This gave us the means to live longer and more productive lives.

It used to be that we didn't have enough resources to provide everyone with enough food, water, and shelter. This limited population growth. Then came industrial farming and mechanized production; humanity finally had the tools to build a world of plenty. At the heart of all this was the discovery of our ability to transform one form of energy to another. The invention of the steam engine allowed us to convert heat to movement. One steam engine could do the work of a dozen horses, only without the kicking and excrement. The fundamental principle at the heart of every steam engine is as follows. A fuel, usually coal, is burnt. The resultant heat boils water to create steam. The steam pushes a piston, which in turn moves anything connected to it.[1]

Steam engines powered factories, trains, and agriculture. In 1880, a coal-fired steam engine was used to power an electric generator, converting energy from movement to electricity. Soon enough, hydroelectric energy made its way onto the scene. Then, petroleum made its grand entrance. This substance, which charlatans peddled as medicine, was quickly firing internal combustion engines. It was only a matter of time until the car replaced the cart as our preferred mode of transport.

At around this time, John Tyndall began to analyze the properties of different gases. He discovered that there were differences in the way that "perfectly colorless and invisible gases and vapors"[2] absorbed and released heat. Tyndall demonstrated that oxygen, hydrogen, and nitrogen didn't absorb heat. He also showed that water vapor, carbon dioxide, and ozone did. Water vapor was the strongest absorber of heat. Tyndall considered these gases to be pivotal in controlling the Earth's surface temperature. Without these gases, he concluded, the Earth would have been "held fast in the iron grip of frost."[3]

To understand the importance of these gases to life on Earth, let us look towards our closest celestial neighbor, the Moon. The Moon has no atmosphere. Its average surface temperature is a chilly -18°C. Without a protective barrier, its surface temperature varies from an inhospitable -170°C at night to a boiling 100°C during the day. The Earth's average surface temperature is a temperate 15°C. The highest-ever recorded temperature on Earth is 71°C while the coldest is -89°C.[4]

Only a few years after Tyndall's experiments, the Swedish scientist Svante Arrhenius became the first person to directly investigate the effects of these gases on the Earth's temperature. In 1895, Arrhenius published a paper called *On The Influence Of Carbonic Acid in the Air upon the Temperature of the Ground*. He argued that concentrations of these gases influenced Earth's surface temperature. As concentrations of carbon dioxide and water vapor went up, so would the temperature of the globe. The opposite would also be true. He even calculated that if levels of atmospheric carbon tripled, the temperature of the Arctic regions would go up by around 8°C or 9°C. In his later book, *Worlds in the Making*, Arrhenius presented what he called the "hot-house theory," a precursor to what we today call the greenhouse effect. He stated that due to "heat protection action of gases contained in the atmosphere" the Earth's surface temperature is 30°C warmer than it would be without these gases. He was startlingly accurate, considering the fact that the difference in surface temperature between the Earth and the Moon is 33°C, the Earth being 15°C and the Moon -18°C.[5] Similarly, Arrhenius' colleague, Arvid Högbom, saw that human-made carbon dioxide emissions, mainly from the burning of coal, would slowly heat up the planet. He believed that this warming would take thousands of years and might even be beneficial to us.

Since the Industrial Revolution, our use of fossil fuels has grown exponentially. Coal, oil, and gas provide approximately 70 percent of the world's energy needs. The American environmentalist Bill McKibben put it best: "They are miracles. A solid and a liquid and a gas that emerge from the ground pretty much ready to use, with their energy highly concentrated. Of the three, oil may be the most miraculous. In many spots on the face of the earth, all you have to do is stick a pipe in the ground and oil comes spurting to the surface. It's compact, it's easily transportable, and it packs an immense amount of energy into a small volume."[6]

Fossil fuels have been a tremendous force for good in the world. To understand this let's create a thought experiment together. Let's imagine that when you wake up tomorrow, all the fossil fuels in the world have disappeared. You crawl out of bed to switch on the lights, but they don't work. You switch on the shower, and nothing comes out of the faucet. Cars, trains, planes, and buses are now useless hunks of metal. There's no Internet. Your phone and laptop still work, but how long will it be until their batteries run out of juice? Food production drops without transport and the necessary technology. We are in the dark, useless and hungry.

At its most fundamental level, the burning of fossil fuels produces the following chemical reaction: hydrocarbon + oxygen = energy + carbon dioxide + water. As fossil fuel use increases, so too will carbon emissions. It is only reasonable to assume that this would result in global warming. But we shouldn't yet rush towards this conclusion. We first need to understand the fundamental science that governs the planet's climate and temperature. We must ask two fundamental questions: Are you and I responsible for global warming? What impacts would a warmer climate have on us? The rest of this chapter will focus on the first question, while the next chapter will focus on the second.

## Energy, Heat & Light

We will come to find that climate change is, in more ways than one, a problem of energy. The scientific definition of energy is the property that gives a physical system the ability to perform work. Energy is a thing that takes many forms that provides everything the ability to do things. The weird thing about energy is that we don't know what it is. We know that it exists in various forms such as heat, kinetic, potential, light, electrical, magnetic, nuclear, and chemical energy. We know that we can't create or destroy it. We know that it can be transferred from one form to another. We may not know what it is exactly, but we do know how to harness it.

We will be looking at many forms of energy throughout this book. In this chapter, our focus is going to be on heat and light. I will begin my explanation of heat by asking you to do something: Please stop moving. Now, if your body temperature hasn't dropped from 37°C all the way to -273°C, called absolute zero as it's the coldest temperature possible, you're still moving. You're made of tiny particles. These particles are always moving. They're forever jiggling and bumping into one another. Heat is the result of this movement. If we look at ice, the particles aren't shaking that much. Because of this lack of movement, water turns into a solid. If we heat up the water, the particles begin to jiggle a lot more. If we make them shake enough, they'll be moving about so much that the water turns to steam. Contrary to our perception, there's no such thing as cold. Cold is just the lack of heat. This is why heat always moves from hot to cold.[4]

Heat is simple enough to understand. Understanding light is a whole other story. You might remember that in *Star Wars: The Empire Strikes Back*, Yoda tells Luke Skywalker, "Luminous beings are we, not this crude matter." He was right. Everything with a temperature above absolute zero gives off light. You being a beacon of light doesn't make you unique. You, a rock, and your neighbor's annoying dog who won't stop barking in the middle of the night are all radiating light. The thing is, we can't see the light. Light, which is electromagnetic radiation, comes is many different forms. We can only see one of them: visible light. The others, like ultraviolet and infrared, are invisible to us. We radiate infrared light, so we can't see the light we're emitting. Describing light can get pretty complicated. It's both a wave and a particle, called a photon, and can be in two places at the same time. For our purposes let's look at light as small bundles of energy called photons. The bundles all move at the same speed but have different wavelengths and frequencies.[7]

## Rebuilding The Planet: A Thought Experiment

Now that we've somewhat wrapped our heads around energy, heat, and light we are ready to partake in a journey of the mind; a thought experiment. You're going to need to use your imagination here. Picture what I say vividly. Imagine the sun in all its grandeur, a giant ball of fire, over a million times the size of our planet, floating in space. Its energy is so immense that its light can be seen millions of light years away. About 150 million kilometers away from the Sun is this little blue ball, our Earth. We often imagine that this ball is just floating through space. It's not. It's orbiting the Sun at a staggering 110,000 kilometers per hour. At the same time, it's turning about its axis at 1,600 kilometers per hour. It takes eight minutes for the light from the Sun to reach the Earth.

Sunlight strikes the Earth and deposits some of this energy as heat to the Earth's surface. The Earth then releases this heat back into space in the form of infrared radiation. The amount of energy coming into the Earth is equal to the amount of energy leaving it. This isn't true, but for the sake of simplicity, let's assume that it is. We'll be looking at the Earth's energy balance later in this chapter so you won't have to wait long to know the right answer.

The amount of energy leaving the Earth in the form of infrared radiation is not equal to the amount of energy coming in from sunlight. Some of the light gets reflected right back into space. Specific objects, like snow, ice, and clouds, act as reflective surfaces. You can think of them as giant mirrors if you'd like. We call this a planet's albedo. Let's compare the albedos of three different worlds. The Earth has an albedo of 30 percent, which means that 30 percent of sunlight is reflected back into space. Venus has a much bigger albedo, 70 percent, because a thick layer of sulfuric acid clouds covers its atmosphere. Mars has a much smaller albedo, 15 percent, because it barely has any cloud cover.[4]

Let's now imagine the Earth as it is today with its lush greenery, cities, deserts, glaciers, clouds, and snowy mountaintops. Now I want you to remove everything from this picture up until you're left with nothing but a solid rock. Because we have no clouds, snow and ice, let's make this a slightly shiny rock so that it still has an albedo of 30 percent. What would the average temperature of the Earth be in such an imaginary scenario? It would be about -20°C. The planet has an average temperature of 15°C. What is the cause of this change in temperature? We've kept the amount of energy the Earth gets from the Sun the same. We've also kept the amount of light reflected back into space the same. The only variable we have changed is the presence of an atmosphere. Instead of an atmosphere, let's put a large glass sphere around the Earth. Now imagine standing on the surface of our imaginary Earth. If we look up, we'll see twenty kilometers above us a glass slab miraculously floating.

Our imaginary glass doesn't absorb any sunlight. But it does absorb infrared radiation. Energy from the Sun in the form of light passes right through the glass and hits the surface of the Earth. When the sunlight hits the ground, the energy that the Earth gains is released back in the form of infrared radiation. This time the infrared radiation doesn't go back into space. The glass absorbs it. The glass directs half the infrared radiation into space. The other half, it directs at the Earth's surface.

The point of this thought experiment isn't accuracy. The point of this exercise is to understand that there must be something surrounding the Earth responsible for regulating its temperature. If we are to calculate the surface temperature of this Earth, we will get about 30°C as opposed to the actual 15°C. Allow me to state the obvious and acknowledge that a glass sphere doesn't surround the world. The surface of the Earth is covered by a layer of gases, which we usually call air, that makes up our atmosphere.

## Greenhouse Gases

Our atmosphere is made up of 78.09 percent nitrogen, 20.95 percent oxygen, 0.93 percent argon and 0.04 percent carbon dioxide. The remainder is a mix of various other gases if we don't include water vapor in the mix. Water vapor usually makes up 1 percent of the mix at sea level, with an average of 0.4 percent throughout the entire atmosphere. We have determined that something in the atmosphere regulates the Earth's temperature. It only stands to reason that either all, one, or many of these gases must be responsible. You'll recall that Svante Arrhenius, in the late 1800s, discovered that certain gases absorb and radiate heat. You'll also recall that Arvid Högbom posited that these gases regulate the Earth's temperature.

So why is it that oxygen and nitrogen don't absorb heat while carbon dioxide, water vapor, and methane do? First, we have to remember that heat is in actuality the vibration or jiggling of particles and atoms. Molecules of oxygen and nitrogen, the two most abundant gases in our atmosphere, are made up of two atoms. These atoms hold on to each other tightly. It is because they bond so tightly that they are unable to jiggle and thus unable to absorb heat. Molecules with three or more atoms are more loosely connected and so can shake and absorb heat.[4] Carbon dioxide has three atoms—one carbon, and two oxygens. Water has two hydrogens and an oxygen. Ozone has three oxygens.

## Factors That Impact Climate

The temperature of the Earth has already risen by 1°C since the beginning of the Industrial Revolution. It has the potential to increase by 6°C by the end of this century. For most of us, temperature variations of 1°C and 6°C don't seem to be substantial. It is not uncommon for the temperature to vary by 10°C or 20°C within a single day. For example, the daytime temperatures in the Sahara Desert can be a scorching 50°C. At nighttime, its temperature can go below a freezing 0°C. So why should we care about 1°C or even 6°C? We often look at temperature changes in relation to weather. Climate and weather are not the same. Imagine that on a beautiful spring day the temperature is a lovely 20°C. Let's say that during the night the temperature drops by 6°C. You feel like 14°C isn't too bad, so you put on a jacket. Now imagine if the entire planet's average surface temperature dropped by 6°C. We call that an ice age.

Weather is the behavior of the atmosphere in the short term. It's usually measured over a period of days or sometimes months. Climate is the behavior of the atmosphere in the long run. It's measured and analyzed over a period of a few years or even a few decades.[8] We usually look at the weather in a more localized context and look at climate on a larger scale. For example, someone living in Paris would only want to know what the weather would be like in Paris. Knowing what the weather's like in Rome wouldn't be very useful to them. But a scientist living in Paris might be looking at trends over the whole of Europe or even the entire planet.

The Earth's climate has changed a lot during its 4.5-billion-year lifespan. There have been periods where the Earth was much hotter than today. Times when jungles covered the polar icecaps. There have also been periods much colder than today. Scientists have hypothesized periods where ice and snow covered the entire planet. Let's put this in context. Humans have only existed for about 200,000 years. We've engaged in heavy industry for just the last three hundred years. We weren't around to cause climate change back then. Why do we think that we're producing it now? To answer this question, we have to analyze the different parameters that influence the Earth's climate. We have to figure out what caused climate change in the past. Here we have to remember that energy is the thing that lets everything do things. If there's no energy in the system, then nothing happens. So where does the climate get all its energy? To state the obvious, it gets it from the Sun. Throughout its lifespan, the Sun's energy output has varied. In fact, the energy radiated by the Sun has increased by about 30 percent within the last 4.5 billion years.[9]

We often imagine that the Earth goes around the Sun in a circle. This isn't wholly true. The Earth orbits the Sun in an ellipsis, an egglike shape. Throughout the age of the Earth, the structure of this ellipsis has fluctuated. Sometimes the ellipsis is close to being completely circular. At those time it's at a constant distance from the Sun year-round. At other times the ellipsis can be more pronounced. When the ellipsis is more pronounced the distance between the Earth and the Sun gets bigger and smaller. When the gap is shorter, and the Earth is closer to the Sun, the Earth's temperature goes up, and vice versa.

We all learned back in primary school that the Earth rotates around an axis. We also learned that its axis being at an angle is the reason we have different seasons. What you might not have learned back in primary school was that the angle of this axis changes over time. It shifts from 22.1° to 24.5° and back again over a 41,000-year period.[10]

When the angle is larger, summers get hotter, and winters get colder.

Climate is not only affected by external forces. Internal dynamics "also play a very important role in determining the variability of the Earth's climate. They can be a direct cause of variability, in the absence of any significant change in the forcing, through interactions between various elements of the system... Important examples are the El Niño Southern Oscillation (ENSO), the North Atlantic Oscillation (NAO) and the Southern Annular Mode (SAM)."[11] However, these patterns are cyclical. While they may cause variations from year to year, they do not cause long-term changes.

## The Theory

The Earth's climate is warming at an unprecedented rate —almost fourteen times faster than the previous world record during a period known as the Paleocene-Eocene Thermal Maximum. The implications to the human race of this trend continuing are catastrophic: more intense storms and increased droughts, cities being under water, and many more end-of-civilization type things. In its purest form, this is the story that the vast majority of climate scientists are selling us today. We are often told that 97 percent of climate scientists agree with this story. But science is not decided by a vote. It's not a democracy. To quote the great Richard Feynman, "It doesn't matter how beautiful your theory is, it doesn't matter how smart you are. If it doesn't agree with experiment, it's wrong." We are left with a few fundamental questions: Is the climate getting warmer? What is the cause? Is this a concern, and if so, to what extent? I will be dedicating the rest of this chapter, and the next, to answering these questions.

## Is The Planet Getting Hotter?

NASA's Goddard Institute for Space Studies (GISS) has analyzed changes in global temperature since the period of 1880. The reason they chose 1880 was that this was the first time when a reasonable amount of meteorological stations were spread throughout the world. Their data shows that global temperatures have risen by about 1°C since the beginning of the Industrial Revolution; see Figure 1. The majority of this warming has taken place in the latter part of the twentieth century and the beginning of the twenty-first century.

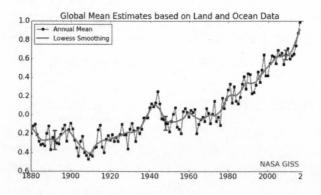

**Figure 1.** *Changes in global temperature from 1880–2017 measured by NASA's GISS.*[12]

We don't only rely on NASA to measure global temperature change. The UK Met Office Hadley Centre and the University of East Anglia's Climate Research Unit in a joint effort, the Nation Oceanic and Atmospheric Administration (NOAA), and the Japan Meteorological Agency (JMA) all independently measure global temperature change. To get this data, the different agencies gather measurements from the air above both land and sea. Sometimes satellite data is also used. When compared to each other, the datasets match pretty well; see Figure 2. Of the four datasets, NASA's GISS is the most detailed. The results are conclusive. The planet is, without doubt, warming at an astronomical rate.

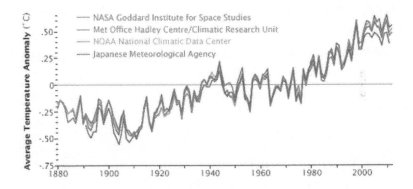

**Figure 2.** *Measurements of global temperature increase from 1880–2012, taken by the four responsible agencies.*[13]

## Determining The Cause

Let's begin to find out the cause of this dramatic warming through a process of elimination. As we learned earlier, the amount of energy the Sun emits influences the temperature of the Earth. You'll note that I said earlier that the amount of energy received by the Earth from the Sun is equal to the amount radiated back into space. You'll also recall that I said that this wasn't actually true. The actual mechanism is this: during periods when the Earth is warming, the amount of energy emitted back into space is less than the amount received. During periods of cooling, the inverse is true.

Ever since 1978 satellites have been used to measure the Sun's energy output, also called solar irradiance. The results gathered by the measurements show that the Sun's total solar irradiance fluctuates up and down by about 0.1 percent every eleven years. The energy emitted by the Sun hasn't increased since 1978. There is no correlation whatsoever between the Earth's observed heating and solar irradiance.[14]

If we turn to look at the changes in the Earth's orbit, we are once again left without an adequate explanation. The Earth's orbit is currently in a phase when it is pretty much close to being a perfect circle. Besides, changes to Earth's orbit occur over the span of thousands of years.[10] What we are currently experiencing is warming over the course of decades. We can discount variation in the Earth's axis of rotation as those changes take thousands of years. What about natural weather patterns? They are cyclical and so cannot be conceivably responsible for such an astounding increase in temperature.[8]

What we do know is that carbon dioxide concentrations in the atmosphere have risen from 280ppm to over 400ppm since the beginning of the Industrial Revolution. Our main energy sources happen to release a lot of carbon into the atmosphere. Or do they? Contrary to what you might think, in the context of nature, we don't emit a lot of carbon.

Natural sources release about 750 billion tons of carbon dioxide a year. All human activity adds a measly 40 billion tons, roughly 5 percent of what nature puts out. The difference is that natural sources also absorb the naturally emitted carbon dioxide to form a balance. We call these sources carbon sinks. Of the measly amount that humans put in the atmosphere, 45 percent gets absorbed by these sinks while the remaining 55 percent stays in the atmosphere. Think of it as a bathtub. If I put water in a bath at a rate of 1 liter per minute and my drainage system can take 1 liter per minute, then all is well. What happens if I decide to increase the amount of water I put into the bath by 5 percent to 1.05 liters per minute? My drainage system still only can take 1 liter per minute. Slowly but surely the tub will fill up, and sooner or later it will overflow.[4]

So how can we be sure that it's us that are emitting this carbon dioxide into the atmosphere? It's not inconceivable that there is another natural process at work. Before we can make such an assumption, we need to analyze a few correlations to find out the actual cause. First, we have to know if there is a correlation between increased carbon dioxide levels and rising temperatures. Lucky for us, we've measured both. Comparing the data, we can see a clear relationship; see Figure 3. Next, we have to find out whether the observed increase in atmospheric carbon dioxide concentrations is due to our emissions. We've measured our carbon emissions and the increase in atmospheric levels. Once again, there's a strong correlation; see Figure 4. To gain further certainty, we have to look at the carbon itself.

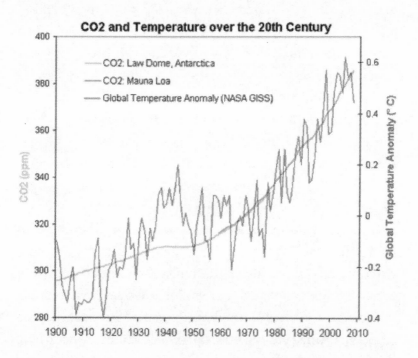

**Figure 3.** *Carbon dioxide emissions against temperature increase.*[15]

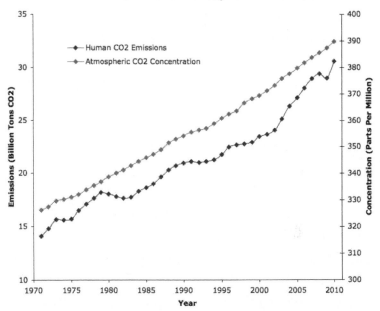

**Figure 4.** *Human carbon dioxide emissions (left y-axis, Source: IEA) vs. atmospheric carbon dioxide concentration (right y-axis, Source: Mauna Loa record).*[17]

You might remember that an ion is an atom that is missing or has extra electrons. An isotope is an atom that is either missing or has extra neutrons. Carbon comes in three isotope varieties. $^{12}C$, in which the atom has six neutrons, is the most abundant form of carbon. $^{13}C$ has seven neutrons; it accounts for about 1 percent of carbon. $^{14}C$ has eight neutrons and is only found in about one out of every trillion atoms of carbon. Unlike $^{12}C$ and $^{13}C$, which are stable, $^{14}C$ is radioactive and dies away after about 50,000 years. Fossil fuels are ancient animal and plant life, millions of years old. They contain much smaller concentrations of $^{14}C$ as compared to the atmosphere. Scientists have observed that as carbon dioxide concentrations increased, the proportion of $^{14}C$ to $^{12}C/^{13}C$ has decreased.

## Climate Change

People usually use the terms "global warming" and "climate change" interchangeably. They are not the same thing. Global warming, which we have analyzed in sufficient detail in this chapter, is the fact that the planet is getting hotter due to our greenhouse gas emissions—nothing more, nothing less. Climate change, on the other hand, refers to the climatic shifts and changes resulting from global warming. Oddly enough, both names find their origins in the same scientific paper, an article published in the journal *Science* in 1975 by Wallace Broecker titled *Climate Change: Are We on the Brink of a Pronounced Global Warming?* To determine the magnitude of the problem we face, we must analyze the impacts that these shifts entail.

## 2

# GAZING INTO THE CRYSTAL BALL

*It's tough to make predictions, especially about the future.*

*— Yogi Berra*

WHEN THE HOUSING bubble burst in 2007, most financial professionals failed to see the crisis brewing. People blamed Wall Street bankers, politicians, mortgage lenders, real estate agents and even journalists. What's perplexing, however, is that most economists, many of whom are paid handsomely to predict what's going to happen to the economy, failed to see the danger right around the corner. Alan Greenspan, revered economist and former chairman of the United States Federal Reserve, believed that it was an unpredictable catastrophe. "Everybody missed it," he said, "academia, the Federal Reserve, all regulators."[1]

It's not just economists who get it wrong. As if by some form of cosmic karma, most predictions about the future tend to be incorrect. In recent years, much has been said about the rise of artificial intelligence, a machine that can think for itself. Some believe we're fifty years away, some think it is twenty years while some believe we're right around the corner. The idea that artificial intelligence is finally upon us is so powerful and tangible to some that they've gone as far as to create a religion based on "the realization, acceptance, and worship of a Godhead based on Artificial Intelligence (AI) developed through computer hardware and software."[2] When one imagines the kind of person who might found such a religion, they might form the mental image of a bearded, fat, middle-aged, balding, paranoid man living in his parents' basement wearing a hat made of aluminum foil. Anthony Levandowski, who founded *Way of the Future*, is no such thing. The engineer, who made his career designing self-driving cars for Google, has a look that I can only describe as uniquely Silicon Valley—a cross between Steve Jobs' bohemian geekiness and Bill Gates' nerdiness mixed in with a helpful serving of Mark Zuckerburg's creepiness. To Levandowski, the rise of the machines is imminent. With great confidence, he once said during an interview, "Machines will be capable, within 20 years, of doing any work a man can do."[3] In another interview, he stated, "within a generation . . . the problem of creating 'artificial intelligence' will substantially be solved."[4] Except, he never said those things. The first quote was by the polymath Herbert Simon in 1956. The second was by the cognitive scientist Marvin Minsky in 1967, thirteen years before Levandowski was born.

The future is a fickle mistress. Her location is forever unknown. The concept of uncertainty is far broader and more encompassing than not knowing what will happen in the future. It is us admitting that we cannot know everything there is to know, that some margin of doubt, big or small, will always be there.

Uncertainty is pervasive. Let us take the simple example of dropping a ball from a height of 1 meter. At what speed would the ball hit the ground? A high school student might calculate, using the basic equations of motion, that it would hit the ground at 4.5 meters per second. A first-year engineering student might estimate the value to be 4.4 meters per second. The variations are admittedly small, but why do they vary in the first place? The high school student assumed acceleration due to gravity to be 10 meters per second squared, while the engineering student believed that the acceleration was 9.81 meters per second squared. The value used by the engineering student is more accurate, but it still doesn't represent acceleration due to gravity with complete accuracy. On average, the acceleration due to gravity on Earth is 9.80665 meters per second squared.[5] So if we put this value into the calculation we should get an entirely accurate result, right? Nope. This is because the Earth's gravitational field isn't uniform. Gravity is stronger in some places and weaker in others. In Mexico City, the acceleration due to gravity is about 9.7760 meters per second squared, while in Helsinki it's 9.825 meters per second squared.[6] So let's say that we're dropping the ball in Mexico City. We now know with certainty what the gravitational acceleration would be. Can we then calculate with certainty the actual speed of the ball hitting the ground? Once again, no. This is because we haven't factored in any forces acting on the ball other than gravity, thus neglecting any forces from air resistance.

Uncertainty exists in everything, even something as simple as calculating how fast a ball will hit the ground. Everyone, scientists most of all, has to come to terms with the concept of uncertainty. But uncertainty doesn't only mean that we do not know things. It also relates to how well we know the things we know. To go back to the ball example, we know with certainty that the ball will hit the ground; we know with certainty that its downward acceleration is primarily due to gravity. There is a small amount of uncertainty, approximately 0.7 percent, about the size of the downward acceleration. There is a more significant, yet still small, uncertainty about how forces due to air resistance will impact the ball's speed. If we were to measure the speed of the ball using a speedometer, there would still be some uncertainty related to the instrument's accuracy.

The mechanisms at play when dropping a ball are simple; thus, the related uncertainties are small, so we consider them negligible and unimportant. But the more complex a system is, the more uncertainty there is related to it. The human mind is the most complex object in the known universe. No wonder there is so much uncertainty in predicting what people will actually do. Predicting what physical objects will do is much more straightforward and has far less doubt attached to it. Even so, a scientist attempting to anticipate the impacts of climate change has their work cut out for them. They must predict the reactions of a complex and interlinking set of physical parameters (the climate) based on the actions of human beings (our emissions).

## A Question Of Energy

Let us look at global warming from a basic principles perspective. Reasoning from basic principles entails boiling things down to their most fundamental truths and then reasoning from there. The basic tenets of global warming are as follows. Concentrations of greenhouse gases in the atmosphere have increased dramatically, due to human activity, in the last two hundred years. Greenhouse gases absorb energy in the form of heat. This has created a global energy imbalance, meaning that the Earth absorbs more energy from the Sun than it emits back into space. This inherently means that the amount of energy inside of the world's climate is increasing and will continue to increase until the amount of energy coming into the Earth and the amount of energy leaving it are equal. Essentially, the world's climate is now more energized than it once was and will continue to get more energized in the future. From this, it would not be unreasonable to assume that increasing global warming will lead to changes in climate patterns. This is backed up by the fact that, in the past, the world was a radically different place when the Earth's average temperature was substantially higher or lower than it is today. The question then becomes, not whether climatic changes will result from elevated global temperature, but instead, what will those changes be and what will be the effect on us?

## The Gold Standard

The best way to know if a specific scientific theory is correct, or at least on the right track, is through practical and reproducible scientific experiments. This is the scientific golden standard. In the usual process that scientific discoveries take, this is step three. Step one is the hypothesis. This, simply put, is the scientist looking out into the universe and making some kind of guess about the way it works. In the second step, the scientist develops the hypothesis utilizing evidence. We call that a theory. Once the theory is formulated, it's verified by experiment. In essence, the majority of science is guesswork that is developed and tested. If the developed guess (theory) matches the observations, it is believed to be correct unless proven otherwise. To see how this works, let's look at Einstein's Theory of Relativity. The basic premise is that time and space are the same thing. You can think of this space-time continuum as a plane. Whenever we put any mass on the plane, the plane bends. It is the bending of this plane which gives us gravity. Einstein also envisioned that the bending of this plane would mean that light did not travel in a straight line but instead is curved by the gravitational pull of another object.[7]

At first glance it might look like Einstein's theory, initially put forward in 1915, is untestable. But two years after it was published, the British astronomer Sir Frank Watson figured out how to test it. He knew that there would be a solar eclipse in 1919. He figured out that the light from the stars in the night sky would bend while passing through the sun's gravity. During the eclipse, the light from the stars would be visible, due to the sun's light not blocking the view. If the positions of the stars changed from their usual place, it would mean that the light emanating from them was bent by the sun's gravity. Sir Arthur Eddington, the person who actually conducted the experiment, traveled to a remote island off the west coast of Africa. There he measured the positions of the stars during the eclipse. He was worried that clouds might cover his view, so he sent another group of scientists to Sobral in Brazil. Lucky for Eddington, the skies were clear in both locations. Did their observations validate Einstein's predictions? The day after the results of the experiment were announced, Einstein was on the front page of newspapers. His radical theory on the nature of the universe was validated.[8]

Unfortunately for us, the only experiment there is to verify the effects of climate change is to continue to do what we are doing and see what happens. We are, in the most literal sense, living inside the experiment. It doesn't take an Einstein to figure out that this isn't a very good idea. To get around this, climate scientists have no other option than to use computer models to predict what the climate of a warmer planet would look like.

The Earth's climate is incredibly complex. To accurately model it, the climate is broken into small pieces. Once we feel that we have a decent understanding of these smaller pieces, we put them together to see how they all play against each other. So how do we figure out if the models are accurate? "When models do a good job replicating past and current events, we assume they are able to forecast, with reasonable certainty, what might happen in the future."[9] Models are, in their most basic form, mathematical representations of all, or at least most, of the factors at play. When we only plug in natural elements into these models, something peculiar happens. While the models represent the climate of the past with relative accuracy, they do an absolutely abysmal job of describing our current environment. It is only after we add in human-induced greenhouse gas emissions that the models, within a reasonable amount of certainty, represent the climate of today.[10]

What is considered the "bible" when it comes to predicting global climate change effects is the Synthesis Report published by the Intergovernmental Panel on Climate Change (IPCC) every five to seven years. The IPCC does not produce any models itself. Rather, it scours scientific literature published in peer-reviewed scientific journals and attempts to present the most common findings of researchers all over the planet. In essence, the IPCC's report presents the most "vanilla" and least controversial views on climate change impacts. These reports have been accused numerous times of being inaccurate. Many have called the projections alarmist and unrepresentative of real observations. This is partly true. The models used by the IPCC have a history of repeatedly missing the mark when it comes to their predictions. However, it is not in the way that most people think. They consistently underestimate real-world impacts. Some examples of their underestimates are: not anticipating significant melting in the Arctic and Greenland ice sheets until 2100 (this is happening now), estimating sea level rise to be 2 mm annually for the early twenty first century (it's double that and rising), and underestimating global carbon sensitivity by 18 percent.[11] All of these inaccuracies are more due to the ever-evolving nature of science; i.e., we know more now than we did twenty years ago, as well as underestimating emissions; i.e., not being able to predict human behavior. The only over-estimate ever found in an IPCC Synthesis Report was the prediction that the Himalayan glaciers would melt by 2035. However, in this case, the authors of the report did not cite peer-reviewed scientific literature. They fessed up to the mistake and released a statement saying, "The IPCC regrets the poor application of well-established IPCC procedures in this instance."[12] Although the IPCC's reports are now considered by most to be the primary source on which policymakers should base their decisions on, the man who first brought the issue of climate change into the consciousness of politicians, and more importantly, the public, was James Hansen.

## Global Heating & The Scorched Earth

In 1988, James Hansen, the director of the NASA Goddard Institute for Space Studies, stood before the United States Senate and said, "I would like to draw three main conclusions. Number one, the earth is warmer in 1988 that at any time in the history of instrumental measurements. Number two the global warming is now large enough that we can ascribe with a high degree of confidence a cause and effect relationship to the greenhouse effect. And number three, our computer climate simulations indicate that the greenhouse effect is already large enough to begin to effect the probability of extreme events such as summer heat waves."[13]

Scientists have understood for a long time that even small amounts of global warming will result in substantial increases in localized heat events, despite uncertainties as to the magnitude, duration, and location of said events. To put this more directly, global warming will, or more accurately is, causing extreme heat waves, but we're not entirely sure where and when they will occur, for how long and how harsh they will be. Within the last century, there was an observed rise in extreme heat waves, with a high increase within the last thirty years.[14] While it is difficult to say whether a single heat event is caused by global warming, we would be blind if we didn't look at the predictions, look at what is happening around us and make the connection.

In June 2015, a searing heat wave struck the southern regions of Pakistan. Temperatures hit 44°C, as compared to the usual 37°C. Two thousand people died from dehydration and heat stroke. The event killed zoo animals and a significant portion of the region's agricultural livestock. Hospitals, morgues, and graveyards became overwhelmed by the sheer influx. Many more would have been killed had it not been for the intervention of the army and several charitable organizations.[15] A year later in 2016, people preemptively dug empty graves in anticipation of another heat wave.[16] In 2017, Pakistan became engulfed in a record-shattering heat wave with temperatures hitting a scorching 51°C.[17]

In 2003, temperatures in Europe were the highest they had been since 1540. This resulted in the deaths of about

2,000 people in Portugal, 1,500 in the Netherlands, 2,000 in the United Kingdom, 300 in Germany, 13,000 in Spain and 15,000 in France. In total, the 2003 European heat wave claimed the lives of over 35,000 people.[18] Fast-forward to 2010, when heat waves rolled over Europe once again, hitting Russia the hardest.[19] Daytime temperatures in Moscow hit a whopping 38°C. One million hectares of land was engulfed in drought-caused fire. In Germany, train passengers collapsed where they stood because the excess heat was causing air-conditioning systems to fail. "You can't attribute isolated events like the heatwaves of 2003 or 2010 to climate change. That said, it's remarkable that these two record summers and three more very hot ones all happened in the last decade. The clustering of record heatwaves within a single decade does make you stop and think," explains Erich Fischer, a researcher at the Institute for Atmospheric and Climate Science at ETH Zurich.[20] This changed in 2016. High temperatures encompassed large swaths of Southeast Asia, southern India, and northern Eurasia, leading to the deaths of at least 580 people in India alone. What makes this heat wave notable is not the heat wave itself, but rather the research that was conducted around it. Researchers set out to figure out how much human-caused global warming contributed to this event. They utilized climate models which factored in both human influences and natural factors as well as models that only factored in natural factors. Their conclusion? "All of the risk of the extremely high temperatures over Asia in 2016 can be attributed to anthropogenic warming."[21] While there have been many events which scientists believe would have been extremely unlikely without human influence, this was the first time when an event was determined to be impossible without human causes.

It is important not to look at extreme heat, as with every other climate change impact, in isolation of other factors such as the socioeconomic situations of the people affected. To understand what I mean by this, let's look at the contribution of climate change to the Arab Spring and the resulting bloody conflict in Syria. The Arab Spring began as a wave of revolutionary protests across a series of Middle Eastern and North African countries, namely: Tunisia, Libya, Egypt, and Syria. The results of these protests were riots, military coups, and civil wars. What were these people demanding? What caused them to flood into the streets?

*Bread, freedom and dignity.*

These nations were heavily reliant on bread as an affordable source of food, with bread being called "aaish" in Egypt and Tunisia, which literally means life. Approximately 40 percent of people living in Egypt and Tunisia live in poverty and are malnourished in some way. Additionally, they are heavily reliant on food imported from other countries. More than half of the food consumed in Egypt is imported. This made these countries vulnerable to shifts in global agricultural markets. The spike in the price of bread, which started in mid-2010, was a direct result of the swath of extreme heat waves hitting the northern hemisphere. These heat waves impacted grain production in such far-off places as Russia, China, and Canada. Egypt was dependent on Russian grain imports to feed its people. Droughts and bushfires, resulting from extreme heat, decimated over a third of Russia's wheat harvest. As a result, Russia placed limits on the amount of wheat to be exported. The decrease in supply, coupled with unchanging demand, caused price hikes. It also wasn't helpful that the government decided to lower food subsidies. "Images of bread became central to the Egyptian protests, from young boys selling kaik, a breakfast bread, to one protester's improvised helmet made from bread loaves taped to his head. Although the Arab revolutions were united under the slogan 'the people want to bring down the regime' not 'the people want more bread,' food was a catalyst."[22]

The Syrian uprising was also catalyzed by environmental factors that would have been extremely unlikely without climate change. Many people believed that Syria was "immune to the Arab Spring." They saw the unrest, which metastasized into a civil-proxy war, as something that came out of nowhere. What many analysts hadn't factored in was "the worst long-term drought and most severe set of crop failures since agricultural civilizations began in the Fertile Crescent many millennia ago," which covered 60 percent of Syrian land between the years of 2006 and 2011. Nearly 75 percent of agricultural crops failed and 85 percent of livestock died.[23] The shocks hitting Syria's agricultural sector impacted the livelihoods of 1.3 million people, with 800,000 people completely losing their income source. A further one million people had limited access to food, or had become "food insecure." The rippling economic tremors threw an estimated three million people into abject poverty. Disenfranchised farmers and herders abandoned their homes in the countryside and embarked on a mass exodus towards the cities, which were themselves overfilled by Iraqi refugees of the 2003 United States invasion. The result was nothing less than a boiling pot of discontent living under the thumb of a despot.[24]

The ability of climate change impacts to multiply already existing socioeconomic problems, in addition to creating threats that weren't there in the first place, is not a controversial stance to take. In fact, the United States Department of Defense, hardly a bunch of hippies singing kumbaya, call it a "threat multiplier" and released a report stating that climate change is "a direct threat to the national security of the United States [that] is impacting stability in areas of the world both where the United States Armed Forces are operating today, and where strategic implications for future conflict exist."[25]

## Water In The Air

It might seem counter-intuitive, but climate change isn't only causing increasing heat waves and drought, it's also increasing rainfall at the same time. When you increase the temperature of the air, its capacity to retain moisture also increases. For every 1°C of warming, the atmosphere holds on to about 7 percent more water. However, this increase in precipitation won't be evenly spread throughout the Earth. While the planet as a whole will, on average, have more moisture in the air, some places will see a lot more rain while others will see a great deal less.[26]

We don't need climate models to tell us this. If we look at the United States, drought caused by elevated temperatures is without a doubt on the rise. In 2012, about 80 percent of agriculture was impacted to some degree by drought,[27] and the amount of extreme heat-related bush-fires has gone up by nearly 400 percent in the last forty years.[28] What about rainfall? Since the late 1950s, precipitation in the southwest has increased by 5 percent, 12 percent in the northwest, 27 percent in the southeast and 71 percent in the northwest.[29] If we look at the southwest of the United States on its own, the region had an average 5 percent increase in rainfall while at the same time being subject to the most severe droughts to hit the area in thousands of years. Australia too has seen noted increases in drought at the same time as increases in rainfall. The same goes for Europe and northeast Asia.

If we were to take the averages of the thirty-nine climate models used by the IPCC, in a 4°C warmer planet, the southern parts of the United States, Mexico, Spain and South Africa are expected to get a 60 percent reduction in precipitation while the southern parts of the Arabian peninsula, parts of North Africa as well as the Arctic and Antarctic are expected to get a 60 percent increase. However, while the models all agree that a warmer world would, on average, be wetter, there is profound disagreement about the distribution of precipitation, i.e., which locations will get wetter and which places will get drier. For example, when all thirty-nine models are averaged out, the prediction is that Australia will see a reduction in precipitation between 5 and 10 percent. However, one of the models in that average predicts that Australia will see a 50 percent drop in rainfall while another predicted a 30 percent increase. We should be careful when looking at climate precipitation models and never rely on the predictions of one single model due to the profound disagreement. This doesn't mean that the models are useless. There are a lot of locations where the vast majority, over 90 percent, of models predict similar changes. There is widespread agreement that the southern parts of the United States, Mexico, northern and southern Africa, southern Europe and southwest Australia will all get drier while India, Bangladesh, Myanmar, north China, Canada, Scandinavia and Russia will all get wetter.[30]

Nevertheless, the models have to improve. Policy makers and city planners will need this data, in the very close future, to make the necessary adjustments to city infrastructure. The reason for this is that the increased precipitation will often manifest itself through extreme deluges which can, in turn, lead to flooding and incredibly expensive infrastructural damage. If the proper infrastructure isn't put in place, areas that become vulnerable to extreme rain could end up having to spend billions of dollars every year on infrastructure maintenance while also suffering from other economic ramifications, such as businesses jumping ship and relocating.

**Food & Water**

Within the last century, the world's population has grown astronomically, from 1.6 billion people in 1900[31] to 7.6 billion people today.[32] The United Nations estimate that the world's population will rise to 9.8 billion in 2050 and 11.2 billion in 2100.[33] At the same time, extreme heat and drought are increasing, thus putting higher pressure on the world's agricultural sector. To make matters worse (unrelated to climate change) an astronomical number of the world's freshwater aquifers have been depleted.[34]

When these factors are taken into account, there's a high chance that growth in the agricultural sector won't be proportional to population growth; i.e., we'll have more mouths to feed with less food. According to the World Bank Group, "Undernutrition is identified as the largest health impact of climate change in the 21st century. A 6 percent decline in global wheat yields and 10 percent decline in rice yields is expected for each additional 1°C rise in global temperature, with substantial impacts on undernutrition and stunting in food insecure or poor regions. An additional 7.5 million children are expected to be stunted by 2030, 4 million of whom are expected to be affected by severe stunting, increasing to 10 million children by 2050."[35]

While food security is something of paramount importance, clean water is without a doubt our most important resource. The average human can survive about three weeks without food; without water, it's less than a week. Over two-thirds of the Earth's surface is covered in water. Unfortunately for us, only 2.5 percent of all that water is fresh water. Most of the world's fresh water is trapped as ice, making it inaccessible. Only 1 percent of the world's fresh water resources are accessible to us.[36] To quote the United States EPA (pre-19th of January 2017), "Water resources are important to both society and ecosystems. We depend on a reliable, clean supply of drinking water to sustain our health. We also need water for agriculture, energy production, navigation, recreation, and manufacturing. Many of these uses put pressure on water resources, stresses that are likely to be exacerbated by climate change."[37] Due to high evaporation caused by excess heat, people's demand for water will increase, while at the same time, elevated evaporation causes fresh water supplies to deplete.

## Acidified Oceans

A third of the carbon we emit ends up in the ocean. When carbon dioxide mixes with water, it forms carbonic acid.[38] This is one of the reasons why soft drinks are so acidic. Oceans are naturally slightly alkaline, with a pH of 8.06. Since the beginning of the industrial revolution, the pH of the oceans has gone down by 0.1. While that might not seem like much, that 0.1 drop in pH means that the oceans are now 30 percent more acidic than they were two centuries ago.[39] The pH scale isn't linear; it's logarithmic. A change in 1 down the pH means that the acidity of a solution has increased tenfold, and an increase of 1 up the pH scale means that a solution is ten times more alkaline.

The current rate of ocean acidification is faster than it has been during any other period within the last 300 million years. The IPCC estimates that, by the end of this century, the pH of the oceans will decline a further 0.3 to 0.4 units,[40] meaning that ocean acidity could increase by over 200 percent by the end of the century.[41] Increased ocean acidity reduces the amount of carbonate in the water.[42] This makes it harder for corals, mollusks, and other shelled organisms to build their shells. Moreover, the decrease of carbonate in the ocean makes calcium carbonate, the material these creatures use to build their shells, more soluble.[40, 43]

Coral reefs alone support about a quarter of all life in the sea. They are among the most diverse ecosystems on the planet, as a collective, rivaled perhaps only by the Amazon rainforest.[44] Not only are they weakened by ocean acidification, but they are also incredibly sensitive to temperature.[45] Warmer, more acidic water has already wreaked havoc on many essential coral reefs around the world, with the most famous example being the Great Barrier Reef in Australia. "Australia's Great Barrier Reef is heralded for its biodiversity: The colorful clusters of coral and wisps of islands stretch 1,400 miles, home to white and orange clown fish, the blacktip shark, humpback whales and hundreds of other species."[46] During 2016 and 2017, half the coral in the Great Barrier Reef died due to the record-breaking heat.[47] Many scientists believe that most of the world's coral reefs won't survive the century, while others think that 90 percent will be dead by 2050.[48] Coral reefs play an integral role in a much broader ecosystem of fish and other marine animals. Over a billion people on this planet rely on food gathered from this ecosystem.[49]

**Storms Brewing**

In 1953, the United States began to name storms alphabetically using female names. By 1978, both male and female names were used to name storms in the Northern Pacific and in 1979 for the entire Atlantic Basin. In 2017, the World Meteorological Organization retired four names from the roster due to the storms being "so deadly or costly that the future use of its name on a different storm would be inappropriate for reasons of sensitivity."[50] The names? Harvey, Irma, Maria, and Nate.[51]

When Hurricane Harvey, a Category 4 storm, struck Texas in August 2017, it resulted in $125 billion in damages, making it costlier than any other storm in U.S. history except Hurricane Katrina in 2005.[52] It produced such unprecedented amounts of rain that the "National Weather Service added two more shades of purple to its rainfall maps to effectively map Hurricane Harvey's rainfall amounts."[53] During the storm's peak, a third of the city of Houston was under water.[54]

Lasting from August 31 until September 11, Hurricane Irma was "the strongest Atlantic basin hurricane ever recorded outside the Gulf of Mexico and the Caribbean Sea."[55] On September 6 it decimated a string of small Caribbean islands in its path. More than 5 million people had to evacuate their homes to escape.[56]

Hurricane Maria, the most intense tropical cyclone of 2017, is considered to be the most devastating natural disaster in Puerto Rico's history.[57] The official number for those killed is placed at sixty-seven. However, the actual death toll is far higher, numbering in the tens of thousands (more on this in a later section).[58] Such devastating events have caused people to ask, are such violent and destructive storms caused by climate change?

It was initially hard to distinguish trends in storm frequency and intensity due to high variation year to year. However, the use of satellites in the 1970s made it possible for us to better track storms. Forty years worth of satellite data shows that there has been an increase in intense hurricane activity.[59] Additionally, there is substantial evidence indicating that typhoons (hurricanes in the western North Pacific Ocean) are also growing more intense.[60] However, the total number of storms has not increased. This is in-line with climate models, the majority of which predict that although the total number of storms will not increase, and may perhaps even slightly decrease, the storms themselves will be more intense and violent.[59] So what are the mechanisms, related to global warming, that can cause storms to become more intense?

The two primary mechanisms are a hotter climate and an increase in sea level (an issue we will tackle in much greater detail in the following section). Warmer sea surface temperatures inherently mean that the water now holds more energy. Hurricanes feed off of this energy, resulting in higher wind speeds. Recall that warmer air can retain more moisture. This leads to higher amounts of rainfall, as was the case with Harvey. Concurrently, an increase in sea level means that when a storm hits a coast-line, the water it brings with it can travel farther inland, leading to far higher coastal flooding.[59] When Hurricane Sandy hit the shores of New York, much of the $65 billion in damage was caused by coastal flooding.[40]

Not only are storms intensifying, but they are also appearing in places and at times that are unexpected. Within recent years, there has been a rapid rise in storms in the Arabian Sea, which sits between Yemen, Oman, and India. Cyclones are rare in the Arabian Sea, "yet in 2014, cyclone Nilofar caused flash-floods in north-east Oman, killing four people. A year later, two cyclones hit back-to-back for the first time. Chapala and Megh both made landfall in Yemen as 'extremely severe cyclonic storms' – with winds as strong as hurricanes – killing 26 people and displacing tens of thousands."[61] In 2018, Cyclone Mekunu brought eight years' worth of rain down upon the Dhofar Governorate of Oman in three days. It was the most intense tropical cyclone to strike the Arabian Peninsula in recorded history.[62]

## Ice Melt & The Rising Sea

Edinburgh, New Orleans, Bangkok, Venice, Miami, Istanbul, Jeddah, Perth, Oslo, Dubai, Manama, Lisbon, Los Angeles, Barcelona, Tokyo, New York—what is it that unites these cities? What do they all have in common? They all lay on, or near to, coasts. Two-thirds of the world's major cities are located, for historical trade reasons, on coasts. It is for this reason that sea level rise is perhaps the most ominous and catastrophic impact of climate change.

Sea level rise during the previous two centuries was modest: six centimeters during the nineteenth century and nineteen centimeters during the twentieth century.[63] Looking at the scientific evidence, the sea level rise we will experience during this century will be anything but modest. To understand why, we have first to appreciate the two basic mechanisms that lead to increases in sea levels.

When we say that the world is now one degree warmer than it was two centuries ago, we are referring to the increase in temperature in the air; however, it is important to remember the oceans are also getting hotter. In fact, "the oceans have absorbed 85 percent of the excess heat trapped by the atmosphere since 1880."[64] As water gets hotter, it expands. This thermal expansion was the primary driver behind the sea level rise experienced during the last two centuries. However, within recent decades, the contribution of thermal expansion has been, in many ways, eclipsed by the melting of land ice (glaciers, ice caps, and ice sheets).[65]

The IPCC estimated that annual sea level rise during the early twenty-first century would be 2 mm per year.[12] It's currently double that value. Why? Because the majority—virtually all—of the models used by the IPCC underestimated or completely neglected the contribution of ice melt to sea level rise.

Models that attempt to simulate the impacts of excess heat on ice sheets, and thus the resultant ice melt, are still at an early stage of development and are poorly understood. As a result, most modelers choose to omit the contribution of ice melt. Odd, considering it is the primary driver. However, rudimentary ice sheet models can present researchers with results that are, quite frankly, hard to take seriously. This is illustrated in the case of a modeler "who found that addition of hydro-fracturing and cliff failure into their ice sheet model increased simulated sea level rise from 2 to 17 m… and accelerated the time for substantial change from several centuries to several decades."[66]

But what if we could bypass the models completely? This is precisely what James Hansen and a team of researchers attempted to do in a 2015 paper published in the scientific journal *Atmospheric Chemistry and Physics*. They used data which showed that approximately 360 billion tons of fresh water was injected into the oceans every year between 2003 and 2015. They then calculated how much sea levels would rise if the rate of ice melt doubled every 5, 10, and 20 years, cutting the models off at the 1- and 5-meter marks. The results of their calculations are troubling at best. If the doubling rate was found to occur every 20 years, a 1-meter rise in sea level would be expected by approximately 2110 and a 5-meter rise close to 2150. This is roughly consistent with the IPCC's current estimates. However, if the doubling rate was found to occur every 10 years, a 1-meter rise in sea level would be expected by approximately 2070 and a 5-meter rise close to 2090. If the doubling rate was found to occur every 5 years, a 1-meter rise in sea level would be expected by approximately 2040 and a 5-meter rise close to 2050.[66] These results have given me more than one sleepless night.

Beginning with the dawn of the Industrial Revolution, two hundred years ago, people have migrated to cities in search of a better life. Since 1950, the number of people living in cities has more than quadrupled, from 746 million to approximately 4 billion. The United Nations estimates that a further 2.5 billion people will live in cities by the middle of this century, due to population growth. This dynamic growth has made humanity more vulnerable to climate change due to rising sea levels and higher storm surges. In 2009, a team of researchers headed by Xingong Li of the University of Kansas utilized geographic information systems (GIS) modeling to attempt to predict the area of land lost and the number of people displaced as the result of a 1-meter increase in sea level as well as a 6-meter increase.[67] The results of their model predicted that 108 million people would be displaced in the event of a 1-meter rise, while 413 million would be displaced at 6 meters. Although far more accurate than previous estimates, a limitation of this study was that it did not factor in population growth. Additionally, the estimates presented by the team, even if population growth is not taken into account, may err on the conservative side as many estimate that a sea level rise of 5 or 6 meters would lead to the complete forced displacement of all those who live in coastal areas with elevations less than 10 meters above sea level, an area referred to as the low elevation coast zone (LECZ).[66]

The LECZ is home to many of our most productive and prosperous cities such as Seoul, Tokyo, Karachi, Jeddah, Bangkok, Singapore and Miami, with the most densely populated areas being in Asia. In 2015, a research team headed by Barbara Neumann of Kiel University published a paper titled *Future Coastal Population Growth and Exposure to Sea-Level Rise and Coastal Flooding – A Global Assessment* which attempted to quantify the total anticipated population growth in the LECZ. Lower forecasts indicated that by 2030 the estimated population of the LECZ would be just under 900 million, while the population in 2060 would be higher than 1 billion. Higher forecasts project that by 2030 the population of the LECZ will be around 950 million and 1.4 billion by 2060.[68] Using such data, Charles Geisler of Cornell University extrapolated the results to indicate that by 2100 the population of the LECZ could be approximately 2 billion.[69]

When we consider how a 5 meter increase in sea level will impact human civilization, in addition to the displacement of potentially billions, we are looking at nothing less than losing total functionality of most, if not all, of our coastal cities.[66] The economic costs would be incalculable. In short, we are looking at the greatest catastrophe in human history. It's all the more frightening when we consider that if the rate of ice melt doubles every five years, we are looking at these events manifesting within many of our lifetimes.

Multi-meter sea level rise is not an unheard of phenomenon within the planet's geological history. The Eemian was an interglacial (warm) period which occurred roughly 130,000 years ago. By analyzing ancient sediments, such as coral reef remains found much higher than sea level, geologists have determined that sea levels were between 6 and 9 meters higher than they are today. There is substantial evidence to suggest that the sea level rise was not slow, occurring on a timescale of centuries, but rapid, happening on a timescale of decades—or, more accurately, within one ecological lifespan.[66]

Data gathered by scientists, such as extracting carbon dioxide samples from ice cores, has shown that greenhouse gas concentrations during the Eemian were lower than they are today. Moreover, the average temperature of the Earth during that period was only a few tenths of a degree hotter than it is today. This is worrying considering that the estimates for the Earth's temperature increase in this century ranges from 2 to 6 degrees, depending on our emissions. It has become commonplace for policymakers to identify 2°C warming as our "safe" limit. However, considering the evidence, 2 degrees might not be safe at all. Looking for evidence in the paleoclimate data is, by necessity, speculative, it is based on proxy data rather than observation. No one was there to see it. To have a more robust understanding of future sea level rise, we must analyze the current condition of the world's ice sheets as they are today.

If I were to ask you to imagine an ice sheet, you might paint a mental image of the North Pole, in the Arctic. You might even choose to put a couple of polar bears in your image. However, the Arctic is not an ice sheet. The North Pole is a large collection of sea ice floating over the Arctic Ocean. During winter, the ice can become 3 or 4 meters thick.[67] However, in some regions where ice has accumulated for many years it can become 20 meters thick.[68] Sea ice is merely ocean water that has been cooled to the point of freezing by the air and, contrary to what many think, it is not something which directly impacts sea level rise. An ice sheet is a large mass of freshwater ice, located on land, taking up an area of more than 50,000 kilometers squared. The earth has two ice sheets: the Greenland ice sheet and the Antarctic (although many scientists classify the Antarctic as being made up of two ice sheets: the Eastern and Western Antarctic ice sheets). "Ice sheets form in areas where snow that falls in winter does not melt entirely over the summer. Over thousands of years, the layers of snow pile up into thick masses of ice, growing thicker and denser as the weight of new snow and ice layers compresses the older layers."[69]

The vast majority, 99 percent, of all the world's fresh water is contained in the Antarctic and Greenland ice sheets. The Greenland ice sheet, which has an area approximately triple that of Texas, contains roughly 1.7 million cubic kilometers of ice, which, if melted completely, would lead to a sea level rise of 6 meters. The Antarctic ice sheet, which has approximately the same area as the United States and Mexico put together, contains roughly 30 million cubic kilometers of ice, enough water to lead to a sea level rise of 60 meters.[67] Satellite imagery has shown that, since the early 1990s, these ice sheets have been melting at an accelerating rate.[70] As the ice melts, fresh water is injected into the oceans, meaning that the oceans have more water in them than they used to have.

The rate at which the Greenland ice sheet gains and/or loses ice is governed by an annual cycle. Throughout June, July and August, Greenland's glaciers lose ice due to elevated summer temperatures. From September through May, the glaciers gain ice. Within normal conditions, the amount of ice lost in the summer is equal to the amount of ice gained throughout the year. However, global warming impacts this cycle in two ways. The increase in atmospheric temperature has accelerated the rate at which the ice melts. However, warmer air is able to retain higher amounts of moisture, meaning that the rate at which snow falls has also increased. The question, then, is: is the amount of increased ice melt counterbalanced by the increase in snowfall? The answer is no. The amount of ice lost due to melt is far higher than the amount gained due to increased snow.[71] The same basic mechanism governs ice melt in Antarctica.

In Greenland, the rate at which the ice melted doubled between 1992 and 2011, indicating a doubling rate of 20 years, with the average contribution of ice melt to sea level being 0.74 mm per year during that time span. However, recent developments have shown that the acceleration rate is, and will continue, to increase within the coming years. Up until the early 2010s, most of Greenland's ice melt came from glacier calving, when ice breaks from the edges of a glacier into the sea. Today, surface melt now injects more meltwater into the oceans than glacier calving.[72]

Within recent years, researchers studying the Greenland ice sheet have noticed strikingly beautiful blue pools of meltwater forming on the surface of the ice sheet. These pools accelerate melt in two ways. The darker color of these pools reduces the ability of the ice sheets to reflect sunlight back into space. This means that the ice sheets absorb a larger amount of heat, increasing the melt rate. As more pools form, the ice sheets absorb even more heat, which results in more melt, which results in more pools, which results in more melt, on and on in a loop.

Additionally, as meltwater makes its way through the cracks and crevices of the ice sheet, it melts more ice and impacts the structural integrity of the ice sheet, making it more likely for glaciers to collapse. Such mechanisms have caused the rate of melt and freshwater injection to double from 2003 to 2017.[72] All of this evidence indicates that the acceleration rate for the Greenland ice sheet melt will increase; essentially, the acceleration rate itself will accelerate.

A study published in the scientific journal *Nature* in 2018, involving eighty-four scientists from forty-four international organizations such as NASA and the European Space Agency, presents us with the most definitive and comprehensive findings and analysis on the health of the Antarctic ice sheets.[73] The researchers found that during the years between 1992 and 2011, the Antarctic lost an average of 83.8 billion tons of ice per year, contributing to an annual sea level rise of 0.2 mm. What is alarming, however, is that the study showed that the rate at which the Antarctic loses ice tripled to 241.4 billion tons between 2012 and 2017; i.e., a *tripling* rate of five years. This is an unprecedented acceleration rate, faster than Hansen's—or indeed any—modeled worst-case scenario.

## Compounding

It is important not to look at climate change risks and impacts in isolation of one another. The 2010 heat wave that struck Russia presents us with an example of how multiple impacts compound to create hazards that are far-reaching, and far more destructive than if they had occurred in isolation of one another. The high temperatures alone are estimated to have killed approximately 11,000 Russians. Additionally, the combination of the heat and drought led to widespread wildfires which decimated much of Russia's agricultural production. "The wildfires also induced large-scale air pollution in cities such as

Moscow, adding to the death toll caused by the heat-wave."[74] An estimated 55,000 people died as a result of the air pollution caused by the fires. And the impacts of the heat wave and drought were not localized to Russia: The reduction in agricultural exports was a direct contributor to political unrest in Egypt and Tunisia in 2011.

We discussed in the previous section how sea level is projected to be a driver for mass migration. It is also crucial for us to understand that climate-change-induced migration will not only be influenced by sea level rise. We've already seen how historic heat waves destroyed much of Syria's agricultural sector, leading to the migration of millions. This was a climate-change-caused mass exodus in which sea level rise played no role. With that, it is impossible for us to predict the magnitude of migration expected due to the sheer number of variables at play.

Many countries are vulnerable to a host of different climate change impacts. For example, in recent years Egypt's production of maize, in the summer, and wheat, in the winter, have suffered lower productivity due to periods of elevated temperatures.[75] Additionally, rising sea levels are affecting the Nile River delta,[76] the area where Egypt grows most of its crops, while high sea temperatures have reduced the nation's fish supply.[77] As temperatures increase, these effects and their related impacts will multiply. An increase of half a meter in sea levels would result in the displacement of 3.8 million people and 1,800 square kilometers of cropland, double the size of Cairo, being flooded and destroyed. A rise of 1 meter would displace 6.1 million people and result in 4,500 square kilometers being underwater.[78] A quarter of

Egyptians work in the agricultural sector. As crop yields deplete, it is not inconceivable that many will be left jobless, just as the case was for Syria. We cannot separate these impacts from pre-existing social, economic and political problems. Much of Egypt is poor, its economy is in tatters, and its population is rapidly increasing. The existing infrastructure will not be able to cope with the projected impacts.

It is imperative that we do not forget our ability to make a bad problem worse. When Hurricane Maria struck Puerto Rico in 2017, the storm itself killed sixty-seven people. It also destroyed much of the country's infrastructure. Many did not have access to clean water and electricity in the months following the hurricane, and at the time that I'm writing this chapter, many still don't. Corporate corruption stalled many initiatives that aimed to provide electricity. The island's medical infrastructure was in tatters. The lack of proper medical resources after the storm caused the deaths of over 4,600 people,[79] deaths that could have been altogether avoided if adequate relief efforts were made.

It is not thought that environmental factors, such as climate change, are direct causes of war and conflict. However, there are many examples of environmental impacts acting as something that puts those in conflict zones at higher risk and something that may contribute to the unrest that causes conflict—the proverbial straw that broke the camel's back, if you will. Although not impacted by climate change, to understand the interplay between environmental factors and violent conflict, let us analyze the case of the Rwandan Genocide.

Most see the Rwandan Genocide as the result of ethnic tensions between the Hutu and the Tutsi peoples. In 1994, between 500,000 and 800,000 (mostly Tutsi) people were killed within a period of less than three months. Between 1990 and 1993, a civil war raged in Rwanda following an invasion of Tutsi refugees from Burundi. In 1993, "a series of negotiations resulted in the adoption of the Arusha accord, which called for the eventual sharing of power between the invaders (known as the Rwandan Patriotic Front or RPF) and the former regime of Hutu President Juvénal Habyarimana, and his party, the Movement for Democracy and Development (MRND)."[80] The massacres, carried out by Hutu militants, began after President Habyarimana's plane was shot down in 1994. However, many believe that ethnic tensions alone cannot explain the bloodshed; the country's environmental and economic situation must also be looked at.

Up until the early 1980s, Rwanda's agricultural practices were, for the most part, sustainable. As population growth increased so did the demand for food. Farmers resorted to unsustainable farming methods which depleted the land, making it unusable for agriculture. Despite the government's efforts, by the late 1980s, much of the country's agricultural land had been depleted. The general condition "in 1993 was such that half of Rwanda's citizens were under 15 years of age. Less than 10% lived in cities; most were living 'up the hill,' in fragile ecosystems that were fast eroding due to deforestation and unsustainable agriculture. Nevertheless, 40% of the gross domestic product was generated by agriculture, there being virtually no other industries in the country."[80] The economic shock of a decimated agricultural sector left its mark on the general population, increasing political tensions and, by extension, ethnic tensions. Separating such environmental and economic factors from the traumatic bloodshed that followed is nothing more than folly.

Although climate change was not a contributor to agricultural degradation in Rwanda during the 1980s and 1990s, the example demonstrates how environmental factors, alone not thought to cause conflict, can exasperate existing problems. The United States Department of Defense profoundly understands this, stating in a 2015 report, "Global climate change will have wide-ranging implications for U.S. national security interests over the foreseeable future because it will aggravate existing problems—such as poverty, social tensions, environmental degradation, ineffectual leadership, and weak political institutions—that threaten domestic stability in a number of countries."[81]

## Feedbacks & Tipping Points

When looking at climate change, we have to differentiate between climate forcings and climate feedbacks. In the past, climate change was initiated by an inciting incident such as an increase in solar irradiance, shifts in the Earth's axis of tilt, and greenhouse gas emissions from volcanic activity or, in our case, greenhouse gas emissions from human activity. These are what we call climate forcings. How reactions in the climate, caused by the forcings, can themselves influence the globe's temperature are called climate feedbacks. For example, during the Paleocene–

Eocene Thermal Maximum, 55 million years ago, global temperatures were far higher than they are today. No one is quite certain of what was the inciting incident that caused the initial warming. Whatever it was, it was enough to cause reactions within the Earth's climate that resulted in more greenhouse gases accumulating in the atmosphere, which in turn caused the majority of the warming.[82] There are two types of feedback: positive and negative. Positive feedbacks are those that exasperate the problem and increase warming, while negative feedbacks are those that help stabilize the climate.

We discussed earlier, in Chapter 1, that the problem of greenhouse gas emissions is a mass balance problem, meaning that we are emitting greenhouse gases at a faster rate than the planet can absorb them. Approximately 55 percent of our carbon emissions stay in the atmosphere. About 30 percent are absorbed by the oceans, and 15 percent by the land. However, the ability of natural processes to absorb our emissions may decrease as time goes on. As temperatures increase, the capacity of the oceans, our biggest carbon sink, to absorb emissions will deplete. To understand this, imagine two cans of Coke. One is warm while the other is cold. Which one is fizzier? The cold one, of course. This is because as the temperature of water increases, the amount of carbon dioxide that can be dissolved in it decreases.[83]

Additionally, there is substantial evidence to suggest that the ability of plants to absorb carbon dioxide will decrease as atmospheric concentrations of carbon increase (we will explore this in greater detail in Chapter 4).[84] Such feedbacks mean that as more of our emissions remain in the atmosphere, due to less natural carbon absorption, we can expect more heating per ton of carbon emitted. This does not only count for our emissions. We discussed in the previous chapter that natural processes emit carbon at far higher rates than we do. However, nature's emissions are absorbed at the same rate which they are emitted, while ours are not. If nature's ability to absorb carbon decreases, more carbon from natural sources will make its way into, and remain in, the atmosphere, resulting in even more warming.

Many believe that global warming has the potential to reach a "tipping point," after which the speed of warming will skyrocket from where it currently is, making any efforts to mitigate it futile. While there is evidence that the feedbacks will magnify global warming, their contribution pales in comparison to our emissions. The two most often cited potential causes for a climate tipping point are greenhouse gas releases from melting permafrost (frozen soil) on land and subsea permafrost. The Earth's atmosphere contains approximately 850 billion tons of carbon. Permafrost, on the other hand, contains approximately 1,400 billion tons of carbon, as well as other greenhouse gases, primarily methane, trapped in ice.[85] If all permafrost melted today, we would be looking at an extinction event on the scale that killed the dinosaurs. Thankfully, however, that is far from the situation we are looking at. The fact that temperatures are rising does not mean that all of the Earth's permafrost will melt. The issue is trying to figure out how much of this soil will thaw, how fast, where, and how much carbon will be released as a result.

A review article written by Carolyn D. Ruppel and John D. Kessler examined the potential for subsea permafrost to cause a climate tipping point. They found no substantial evidence for such a view.[86] According to their findings, "Methane bubbles emitted at water depths greater than about 100 m are unlikely to retain their methane as they ascend to the sea-air interface. The methane that dissolves in the water column is often oxidized to carbon dioxide through microbial processes. Seafloor methane emissions are not entirely benign for ocean chemistry, but gas hydrate degradation likely makes insignificant contributions to global atmospheric methane concentrations."[87]

A seven-year research project conducted by Christian Knoblauch from Universität Hamburg found that land permafrost was melting at a rate faster than anticipated. Based on this research, it is anticipated that "The permafrost soils of Northern Europe, Northern Asia and North America could produce up to one gigaton (a billion tons) of methane and 37 gigatons of carbon dioxide by 2100."[88] While these figures are substantial, we must look at them in their proper context. We current emit approximately the equivalent of 40 billion tons of carbon dioxide annually, meaning that emissions from permafrost this century will be the equivalent of less than two years of human-caused emissions.

While feedbacks will contribute to global warming, their total contribution pales in comparison to that of man-made emissions. Feedbacks may add a few tenths of a degree of warming this century; however, if we continue to emit the way we currently do, we are looking at a global temperature increase of 2°C compared to preindustrial levels by 2036 (we will explore this in more detail in Chapter 4). It will primarily be our actions, not natural feedbacks, that determine what the Earth's future climate will look like.

## Inevitable?

Looking at the impacts of climate change gives many a sense of existential dread—and for good reason: these impacts are frightening. However, we have to understand that while climate change impacts have already begun to manifest, the more extreme implications are still not yet upon us and will only come to pass if we continue to emit greenhouse gases at the rate we are currently emitting them today. That being said, the technologies needed to shift us towards a carbon-free future already exist and must be deployed with great haste if we are to mitigate our impacts on the globe's average temperature. The scale and magnitude of climate change's impacts on humanity will be directly proportional to the scale and magnitude of our actions within the coming years. Reducing our greenhouse gas emissions and moving towards more sustainable practices is, without a doubt, humanity's most important mission this century.

# CONSPIRACY, DECEIT & DEFAMATION OF CHARACTER

*It is only by a fierce defense of the truth and a common set of facts that we create the conditions for a democratic free society.*
— Rex Tillerson

THE EVIDENCE for global warming and climate change is overwhelming. Considering the overbearing weight of proof, we have to ask, why is there a climate change debate in the first place? The simple fact is, there is a dark conspiracy at the very heart of this debate.

James F. Black's words to the management committee of the oil giant Exxon were blunt and clear, "In the first place, there is general scientific agreement that the most likely manner in which mankind is influencing the global climate is through carbon dioxide release from the burning of fossil fuels." This was 1977, and the management committee was made up "of the top level senior managers at Exxon. The chairman, the president, the senior vice presidents, corporate wide." It would only be a year later that Black, who was a high-ranking technical expert in the company's Research & Engineering division, would bring his message to a wider audience. He informed Exxon's engineers, scientists, and managers that a doubling of atmospheric carbon dioxide would increase the globe's temperature by about 1.5°C to 4.5°C, with the most likely estimate being 3°C, by the end of the twenty-first century. This was not mere observation, Black firmly understood the enormous risks ahead, "Some countries would benefit but others would have their agricultural output reduced or destroyed."[1]

The Exxon of the 1970s was a company on the very cutting edge of research. It was a company that analyzed all risks to its bottom line (whatever they might be, environmental or otherwise) and took quick, decisive but well-thought-out and informed action to mitigate those risks. In 1977, Exxon's chairman, Clifton Garvin, hired Edward David to head Exxon Research. Garvin wanted to push the company towards greater innovation and saw hiring David, who had spent more than twenty years working at Bell Labs, a research institution known for its technological innovation, as a way to do just that. During those twenty years, he would go on to become Bell Labs' director of research as well as being Richard Nixon's science advisor between 1970 and 1973. David first began to learn about climate science during his time at the White House. Later at Exxon, he opened the floodgates. But it wouldn't be David who would become the hero of Exxon's carbon dioxide research program. That honor belongs to Henry Shaw, the man who would go on to head the initiative.

"The rationale for Exxon's involvement and commitment of funds and personnel is based on our need to assess the possible impact of the greenhouse effect on Exxon business," wrote Shaw to David, at the time his boss. "Exxon must develop a credible scientific team that can critically evaluate the information generated on the subject and be able to carry bad news, if any, to the corporation. We see no better method to acquire the necessary reputation than by attacking one of the major uncertainties in the global carbon balance, i.e., the flux to the oceans and providing the necessary data."[2]

Researchers at the time knew that the oceans could absorb carbon dioxide. The problem was they didn't know how much of our carbon emissions would end up there. They reasoned that if all, or most, of our emissions, ended up in the depths of the sea, the problem would be out of the way for at least a few centuries. If this were the case, it would give the oil industry a lot more wiggle room and allow them to, to quote a former Governor of Alaska, "burn baby, burn." If it turned out that the oceans had a limited capacity to store carbon, the situation would be, by necessity, more complicated. To tackle this issue, Exxon's researchers trekked towards the oceans. They retrofitted one of the biggest supertankers in the world, called the Esso Atlantic, with a cutting-edge lab and planned to collect carbon dioxide samples from the water and the air during its voyage from the Gulf of Mexico to the Arabian Gulf.

"The tanker project was intended to provide valid,

legitimate, scientific data, unassailable hopefully, on key questions in atmospheric chemistry [of] $CO_2$ emissions," said Richard Werthamer, Henry Shaw's boss between 1980 and 1981, "Henry's additional goal was to make Exxon a credible participant in that research and in the dialogue that would inevitably follow... He wanted Exxon to be respected as a valid player and have Exxon's opinions solicited, and participate in discussions on policy, rather than have the issue suddenly dumped with Exxon's back turned."[2]

In 1979, a vice president at Exxon hired Steve Knisely to determine how a warming planet might impact fuel use. Knisely's rudimentary climate models showed that if we continued down the business as usual path, the result would be "noticeable temperature changes" with carbon dioxide concentrations hitting 400ppm by 2010. He was only off by a few years. "The potential problem is great and urgent," wrote Knisely. Regarding the impact on Exxon's bottom line, his warnings were dire. He estimated that we would need to keep 80 percent of recoverable fossil fuel reserves untouched if we were to stop atmospheric carbon concentrations from doubling. However, he also noted that, "Too little is known at this time to recommend a major U.S. or worldwide change in energy type usage but it is very clear that immediate research is necessary." Knisely's research, along with a significant cut in federal funding, drove the company to trash the tanker project.[3]

Instead, it threw its might into advanced computer climate models. By 1981, virtually all of Exxon's scientists came to the same conclusion. If greenhouse gas emissions were not constrained, the resultant global warming would yield catastrophic impacts on hundreds of millions, if not billions, of people worldwide. Exxon's director of the Theoretical and Mathematical Sciences Laboratory at Exxon Research stated that it was "distinctly possible" that the warming, post-2030, "will indeed be catastrophic (at least for a substantial fraction of the earth's population). This is because the global ecosystem in 2030 might still be in a transient, headed for much more significant effects after time lags perhaps of the order of decades."[3] He knew that even if there were no catastrophic impacts by 2030, we would have accumulated enough carbon in the air to make it virtually certain that catastrophe would follow in the subsequent decades.

> *Over the past several years a clear scientific consensus has emerged regarding the expected climatic effects of increased atmospheric $CO_2$. The consensus is that a doubling of atmospheric $CO_2$ from its pre-industrial revolution value would result in an average global temperature rise of $(3.0 \pm 1.5)°C$. There is unanimous agreement in the scientific community that a temperature increase of this magnitude would bring about significant changes in the earth's climate, including rainfall distribution and alterations in the biosphere.*

At the time, the vast majority of people had no idea what global warming, climate change or the greenhouse effect were. It would not be the voice of Exxon, or indeed any other major corporation, that would bring this issue to the masses. It would be a mild-mannered, soft-spoken man from southwest Iowa named James Hansen. "Global Warming Has Begun, Expert Tells Senate," read the *New York Times* headline.

> *The earth has been warmer in the first five months of this year than in any comparable period since measurements began 130 years ago, and the higher temperatures can now be attributed to a long-expected global warming trend linked to pollution, a space agency scientist reported today.*
>
> *Until now, scientists have been cautious about attributing rising global temperatures of recent years to the predicted global warming caused by pollutants in the atmosphere, known as the 'greenhouse effect.' But today Dr. James E. Hansen of the National Aeronautics and Space Administration told a Congressional committee that it was 99 percent certain that the warming trend was not a natural variation but was caused by a buildup of carbon dioxide and other artificial gases in the atmosphere.[4]*

The cat was out of the bag. Exxon knew the ramifications to their business. Rather than disclosing that it had a deep understanding of climate change, it denied it completely. Despite their research, they argued that there was no link between carbon emissions and climate change. "They spent so much money and they were the only company that did this kind of research as far as I know," said Edward Garvey, the man who worked tirelessly alongside Henry Shaw, in a recent interview. "That was an opportunity not just to get a place at the table, but to lead, in many respects, some of the discussion. And the fact that they chose not to do that into the future is a sad point."[1]

However, *InsideClimate News*, the Pulitzer Prize-winning news organization who had unearthed the internal documents related to Exxon's carbon research program, have also found that Exxon was not the only fossil fuel company that had known about climate change. The American Petroleum Institute, Exxon, Mobil (Exxon and Mobil would later merge to become ExxonMobil), Amoco, Phillips, Texaco, Shell, Sunoco, Sohio as well as Standard Oil of California and Gulf Oil had all shared research concerning climate change and carbon emissions between 1979 and 1983.[5] All of these companies shared, what we might call today, a "conventional" view of climate change. Despite this, together they led a heavily funded campaign with the explicit purpose of sowing doubt around the science of global warming and climate change. The strategy they used was neither new nor unique. They utilized the same approach used earlier by big tobacco to sow doubt about the fact that cigarettes cause cancer.[6]

During the 1950s, evidence began to pile up linking smoking to cancer. This prompted the tobacco industry to found the Tobacco Industry Research Committee (it would later be renamed the U.S. Tobacco Institute) with the express intent of researching the link. The Committee's studies resulted in the publication of the *Frank Statement to Cigarette Smokers*. Within this document, Big Tobacco questioned the validity and authenticity of emerging science while at the same time promising that they would conduct more research to analyze the safety of their product.[7] Instead of conducting research, Big Tobacco hired a public relations (PR) firm called Hill & Knowlton.[6] Hill & Knowlton knew that it was difficult, if not impossible, for the companies to outright deny what the researchers and scientists were saying. But what they also realized was that what they could do was shake the public's confidence in these scientists. They created doubt around anything and everything that linked cigarette smoking to bad health, "even the most indisputable scientific evidence." They used and even secretly created, organizations that would seem to be completely independent, but that were in reality just mouthpieces for Big Tobacco. They created debates designed to bring into question scientific truths that had long been accepted as fact. They also lobbied the hell out of government officials including members of Congress in the United States. Big Tobacco knew that they didn't have to prove that tobacco was safe to fool a large number of the general public, as well as those in positions of power. All they had to do was "maintain doubt," and victory would be theirs.

"Doubt is our product, since it is the best means of competing with the 'body of fact' that exists in the minds of the general public. It is also the means of establishing a controversy,"[8] said a now infamous internal memo from Brown & Williamson tobacco. To take just one example of their nefarious deeds, tobacco companies led a campaign to undermine the World Health Organization (WHO) from the inside. "They paid WHO employees to spread misinformation, hired institutions and individuals to discredit the international organization, secretly funded reports designed to distort scientific studies, and even covertly monitored WHO meetings and conferences."[6]

Just as Big Tobacco founded the Tobacco Industry Research Committee, Exxon, the American Petroleum Institute, and a host of other fossil fuel companies, car manufacturers, and other industrial institutions and corporations founded the Global Climate Coalition in 1989, with the sole aim of stifling public awareness concerning climate change.[9] Utilizing a handpicked group of scientists, the Global Climate Coalition set about bringing attention to, and magnifying, the uncertainties of climate science, as well as fabricating uncertainty out of thin air. The Coalition launched a direct campaign against climate scientists as well as nearly any research conducted on the issue of climate change.

The Western Fuels Association, an alliance of coal producers, suppliers, and utilities, collectively worth over $400 million, produced a half hour long film called *The Greening of Planet Earth,* which made the claim that excess carbon dioxide in the atmosphere would make plants grow faster and stronger and so would be a boon for the world's agriculture. As such, the film argued, any attempt to lower carbon emissions would be misguided and a detriment to us and the planet. The Global Climate Coalition sent the film to thousands of journalists as well as the White House. There it gained popularity and profoundly influenced the decisions of George H. W. Bush, President at the time, as well as the then-chief of staff, John H. Sununu, ahead of the 1992 Earth Summit in Rio de Janeiro, Brazil. The film was narrated by Sherwood Idso, the head of the Center for the Study of Carbon Dioxide and Global Change, an Exxon funded climate change denial vehicle.[10]

However by the late 1990s, as the evidence for global warming began to pile up, many companies, including major fossil fuel players such as British Petroleum, Texaco, and Shell pulled out of the Global Climate Coalition, faced facts and accepted the evidence. However, Exxon, the same company that possessed the greatest knowledge on climate change and its ramifications, stayed the course.[6]

At around the same time, in 1997 to be precise, the world's major industrialized nations had committed to reducing their greenhouse gas emissions through an international treaty called the Kyoto Protocol (more on this in Chapter 5). Now there was a newer and more explicit mission. Destroy the Kyoto Protocol, and if that wasn't possible, get the United States to pull out. In 1998, the Global Climate Science Communications Plan was created by the Global Climate Science Communications Team, which in itself was organized and led by the American Petroleum Institute along with Exxon (no surprises here), Chevron, the Southern Company as well as a litany of conservative "think tanks" such as the Committee For A Constructive Tomorrow, the Marshall Institute, Frontiers of Freedom, and Americans for Tax Reform. The purpose of the "Plan" was simply to misinform the public, the media and politicians to such an extent as to stagnate and end all climate-related action. An internal memo written by the American Petroleum Institute's Joe Walker states that

*Unless 'climate change' becomes a non-issue, meaning that the Kyoto proposal is defeated and there are no further initiatives to thwart the threat of climate change, there may be no moment when we can declare victory for our efforts. It will be necessary to establish measurements for the science effort to track progress toward achieving the goal and strategic success.[11]*

Walker's leaked Orwellian memo also states:[12]

<u>*Victory Will Be Achieved When:*</u>

- *Average citizens "understand" (recognize) uncertainties in climate science; recognition of uncertainties becomes part of the "conventional wisdom."*
- *Media "understands" (recognizes) uncertainties in climate science*
- *Media coverage reflects balance on climate science and recognition of the validity of viewpoints that challenge the current "conventional wisdom."*
- *Industry senior leadership understands uncertainties in climate science, making them stronger ambassadors to those who shape climate policy*
- *Those promoting the Kyoto treaty on the basis of extent science appears to be out of touch with reality.*

Within these spine-chilling words lies one, and only one, true statement. "Industry senior leadership understands uncertainties in climate science." Industry senior leadership did understand the science of climate change, intimately. They knew its ramifications and the cost to human society. They knew that it would bring untold pain and misery to the lives of billions. They didn't care.

The Global Climate Coalition, which was only one of many different initiatives created by Exxon and the American Petroleum Institute, spent a total of $13 million to combat the Kyoto Protocol. When, in 2001, the United States pulled out of the Kyoto Protocol, the Global Climate Coalition had considered itself the victor and disbanded. Without the United States, the impact of the Protocol was hindered. Although this might appear to be the end of the story, many at the time thought it was, many fossil fuel corporations such as ExxonMobil, Peabody Energy, and the Koch Brothers, although most assuredly not all, had a new strategy called "information laundering."[6]

They knew that it would be hard, or more accurately downright impossible, to gain the public's trust when it came to the climate change issue. As such, fossil fuel companies began to fund, and as a result inject their viewpoints into, long established and seemingly independent nonprofit organizations. These include but are not limited to the American Enterprise Institute, the Competitive Enterprise Institute, the Cato Institute, the American Council for Capital Formation Center for Policy Research, the American Legislative Exchange Council, the Committee for a Constructive Tomorrow, and the International Policy Network.[13] Additionally, fossil fuel companies and the institutions they fund began to "secretly pay academics at leading American universities to write research that sows doubt about climate science and promotes the companies' commercial interests."[14]

To expose this campaign of misinformation, reporters, working on behalf of Greenpeace, had gone undercover "claiming to be representatives of unnamed fossil fuel companies." A professor (whose name I have chosen to omit as an undeserved courtesy) among those approached by the reporters was asked to produce a report that would cast doubt on the fact that coal usage is linked to millions of premature deaths, "in particular the World Health Organization's figure that 3.7 million people die per year from fossil fuel pollution." He accepted the task and told the reporters that a full report would cost them $15,000, while a newspaper-style article would cost $6,000. He also informed the reporters that, "There is no requirement to declare source funding in the US." Chief among the professor's clients and financiers were Peabody Energy, a coal utility, who had once paid him $50,000 for a report. The reporters also approached another professor (whose name I have also chosen to omit as an undeserved courtesy. He informed the reporters that he was willing to develop "research" that would detail the benefits of carbon emissions (funny how the research's conclusion came before the actual research). Even though he called his "research" a "labor of love," his rate for doing the work was $250 per hour. The money was to be given to the $CO_2$ Coalition, a climate change denial organization he was a board member of. When asked whether he had to disclose the source of funding for the research, he replied, "If I write the paper alone, I don't think there would be any problem stating that 'the author received no financial compensation for this essay.'"[15]

Perhaps more sinisterly, the misinformation campaign led to a concerted attack on climate scientists. In 2009, a server containing the emails of leading climate change researchers at the Climatic Research Unit at the University of East Anglia was hacked. The compromised content included over a thousand emails and over two thousand researcher documents.[16] After combing through the mountain of emails only two (seemingly) incriminating phrases were found:

*1. The fact is that we can't account for the lack of warming at the moment and it is a travesty that we can't.*

*2. Mike's Nature trick of adding in the real temps to each series for the last 20 years (i.e., from 1981 onwards) and from 1961 for Keith's to hide the decline.*

The phrases used in these emails, when taken out of context, provide a seemingly perfect story of corruption and conspiracy. The Climatic Research Unit wanted to show that the earth was indeed warming but looking at the data they couldn't "account for the lack of warming," which was a "travesty." To fool the public, the researchers used "Mike's Nature trick" to "hide the decline" in temperature and maintain the global warming narrative. The timing of the hack was planned to occur close to the Copenhagen climate talks. The hacks led to an increase in climate change skepticism within the general public. The Copenhagen climate talks were an unmitigated disaster (more on this in a later chapter), and while "Climategate" was not the only factor that led to the talk's failure, it was undoubtedly a major contributing factor. But what narrative did the emails present if read within their proper context?

The researchers were referring to the, now famous, "hockey stick" graph. The graph—created by Michael Mann, the director of the Earth System Science Center at Pennsylvania State University—utilized various sources to present the average change in global temperature over a thousand year period. Tree ring data was used to give an approximation of the earth's historical temperature changes. This is because the rate at which trees grow correlates with temperature and we can gauge the growth rate by analyzing tree rings. When comparing temperature data gathered from thermometers with temperature data gathered from tree rings we see a high correlation from the 1880s up until the 1960s. The problem is that in many high latitude sites, the results began to diverge in the 1960s. However, tree ring data gathered from southern latitude sites still show a strong correlation. Unfortunately, the information about tree ring data in southern latitudes was published in 2004 while the "hockey stick" graph was published in 1998. Scientists had long known about the tree ring problem. Keith Briffa, one of the CRU researchers whose emails were hacked, had already published a paper on the issue in the late 1990s. "Mike's Nature trick" was to use temperature data gathered from thermometer readings, which are far more accurate than tree-ring data, to present the earth's average temperature from 1960 onwards for the hockey stick graph published in the journal *Nature*. The phrase "hide the decline" refers to the data given by tree ring data which, as we've already discussed, was unreliable post-1960. The phrase "we can't account for the lack of warming at the moment" was a reference to the researchers not knowing why, at the time, the reason for the divergence, which was a "travesty."

With all of this said, we have to ask the question, how is climate change perceived by the general public worldwide? Moreover, we also have to ask whether the denial campaign has played a significant role in determining public opinion? Between 2007 and 2008, the Gallup Organization surveyed people from 128 countries to develop the first comprehensive study of climate change opinions worldwide.[17] The results of the study estimate that 61 percent of people worldwide have heard of global warming and 47 percent believed that it is a threat. Awareness was found to be higher in developed nations as opposed to their developing counterparts. The countries with the most robust appreciation of the issue were found to be in Latin America, followed closely by developed countries in Asia such as Japan, where 99 percent of the population was aware of climate change.

While awareness often translates to concern, this was not always found to be the case. For example, in the United States, 97 percent of the population was aware of climate change, but only 49 percent believed it was caused by human activity. In the United Kingdom, 97 percent of the population was aware of climate change, just 48 percent thought it was caused by human activity. In Russia, 85 percent of the population was aware of climate change, only 52 percent believed it was caused by human activity. It might be expected that within nations that have been already severely impacted by climate change, awareness and concern would also be high. This isn't the case. In Syria, only 56 percent of the population were aware of climate change, 54 percent believed it was caused by human activity, and 41 percent think it to be a threat. In Pakistan, only 34 percent of the population was aware of climate change, just 25 percent thought it was caused by human activity and 24 percent believe it to be a threat. While these results do indicate a low awareness of the problem, they also suggest that those who are aware also share a high degree of concern. It is difficult to quantify precisely how much the denial machine has impacted awareness and perception, but it is worth noting that within the country in which the machine is strongest, the United States, there is a high disparity between awareness and concern.

The denial machine is not only concerned with manipulating public opinion. It also aims to bypass public opinion entirely through the manipulation of political actors and decision makers. Nowhere is this more apparent than the Republican Party in the United States. For example, during the 2006 elections for the governor of Alaska, incumbent Democrat Tony Knowles stated that

> *Scientific evidence shows many areas of Alaska are experiencing a warming trend. Many experts predict that Alaska, along with our northern latitude neighbors, will continue to warm at a faster pace than any other state, and the warming will continue for decades. Climate change is not just an environmental issue. It is also a social, cultural, and*

*economic issue important to all Alaskans. As a result of this warming, coastal erosion, thawing permafrost, retreating sea ice, record forest fires, and other changes are affecting, and will continue to affect, the lifestyles and livelihoods of Alaskans. Alaska needs a strategy to identify and mitigate potential impacts of climate change and to guide its efforts in evaluating and addressing known or suspected causes of climate change.*

Later during the election campaign, he would go on to propose

*...the prioritization of climate change research in Alaska; development of an action plan addressing climate change impacts on coastal and other vulnerable communities in Alaska; policies and measures to reduce the likelihood or magnitude of damage to infrastructure in Alaska from the effects of climate change; the potential benefits of Alaska participating in regional, national, and international climate policy agreements and greenhouse gas registries; the opportunities to reduce greenhouse gas emissions; and the opportunities for Alaska to participate in carbon-trading markets, including the offering of carbon sequestration.*

His election rival, Sarah Palin, a Republican who would go on to win the election retorted, "Global warming my gluteus maximus." Except, Knowles never said anything of the sort. All three quotes belong to Sarah Palin. The first and second quotes were from a 2007 Administrative Order issued by Palin, which established the Alaska Climate Change Sub-Cabinet.[18,19] The third is a tweet from 2013.[19]

Palin's shift in opinion closely follows the trajectory of her entire political party, a marked swing in Republican positions regarding climate change during the Obama era. Two years after Palin became the governor of Alaska, she found herself as the vice presidential running mate to presidential hopeful John McCain during his run against Barack Obama in 2008. One of McCain's campaign ads contained, "Rapid-fire images of belching smokestacks and melting ice sheets... followed by a soothing narrator who praised a candidate who had stood up to President George W. Bush and 'sounded the alarm on global warming.'"[20] While George W. Bush was by no means a champion for climate change action, it is important to note that he did not deny the issue completely. When asked, during a presidential debate with Al Gore, whether he was willing to create and implement laws that do something about it, replied, "Sure, absolutely, so long as they're based on science, and they're reasonable." He went on to call global warming "an issue we need to take very seriously."[21] Twenty years earlier, when the issue of global warming began making its way into the public consciousness, his father, then-president George H. W. Bush, stated, "Those who think we are powerless to do anything about the greenhouse effect forget about the 'White House effect'; as President, I intend to do something about it."[22] Mitt Romney, the presidential candidate frontrunner for the Republican Party during Obama's second term, stated that it is "important for us to reduce our emissions of pollutants and greenhouse gases that may well be significant contributors to the climate change and the global warming that you're seeing."[23] Up until recently, Republican voters had the chance of voting for a candidate who pledged to act to combat climate change.

Today, this is no longer the case. Most Republican candidates flat-out deny either the existence of climate change or the fact that it is man-made. All candidates who do claim to believe in climate change, nevertheless argue that they would do nothing about it. This has left Republican voters with no choice but to vote for someone who either denies climate change or pledges to do nothing about it, in practice the same thing. When asked about the issue, Republican presidential candidate Jeb Bush stated, "I don't think the science is clear on what percentage is man-made and what percentage is natural. It's convoluted. And for the people to say the science is decided on this is just really arrogant." Former Director of Pediatric Neurosurgery at Johns Hopkins Hospital, Ben Carson, stated, "I'll tell you what I think about climate change. The temperature's either going up or down at any point in time, so it really is not a big deal." Chris Christie, while saying that he admits climate change is real, went on to add, "The degree to which it (carbon emissions) contributes to it is what we need to have a discussion about." He called for a "global solution" to climate change while opposing all programs that limit emissions. Ted Cruz, at an event sponsored by the Koch brothers, known funders for climate change denial organizations, stated succinctly, "If you look to the satellite data in the last 18 years there has been zero recorded warming (see Chapter 1). Now the global warming alarmists, that's a problem for their theories. Their computer models show massive warming the satellite says it ain't happening. We've discovered that NOAA, the federal government agencies are cooking the books." Carly Fiorina stated that she believes that there is a scientific consensus on the matter, but we still shouldn't do anything about it. George Pataki, who spent the majority of his political career advocating for emissions reduction, refused to comment on his stance and policy opinions the moment he announced his presidential candidacy. Rick Perry called global warming a "contrived, phony mess." During an interview with CBS Marco Rubio stated, "I believe climate is changing because there's never been a moment where the climate is not changing." Rick Santorum, during a Fox News interview, said, "Any time you hear a scientist say the science is settled, that's political science, not real science, because no scientists in their right mind would say ever the science is settled." And Donald Trump, the man who won the election? "This very expensive GLOBAL WARMING bullshit has got to stop. Our planet is freezing, record low temps, and our GW scientists are stuck in ice."[24]

Not a single serious contender, Republican or Democrat, during the 2016 presidential race, did not receive some funding from fossil fuel interests. Hillary Clinton's campaign received over $300,000 from oil and gas interests.[25] However, the vast majority of the funding ended up on the Republican side. An analysis of Federal Election Committee data compiled by Greenpeace found that fossil fuel interests injected more than $100 million into the Republican presidential campaign, meaning that the fossil fuel industry is the single largest backer of the Republican party.[26] Donald Trump's inauguration itself was heavily funded by fossil fuel heavyweights, including "Chevron ($525,000), Citgo ($500,000), ExxonMobil ($500,000), BP Corporation of North America ($500,000), and coal mining company Murray Energy ($300,000)."[27]

During his campaign, Trump pledged to withdraw the US from the Paris Climate Accords (we'll explore the details of the agreement in a later chapter). On the 1st of June 2017, he announced just that.[28] The deal was initially signed by every country on the planet, except for Syria, who's at war, and Nicaragua, who felt that the deal wasn't enough and is planning to be 90 percent reliant on renewable energy by 2020. Nicaragua has since signed, and Syria has shown interest in doing so. The announcement was met with criticism worldwide, with dissenting voices ranging from Greenpeace to the European Union to academics such as Neal DeGrasse Tyson and Noam Chomsky[29] to celebrities like Arnold Schwarzenegger[30] and Tom Morello[31] to blue-chip companies like Facebook, Apple, and Microsoft.[32] To their credit, major fossil fuel giants ExxonMobil, BP, Chevron, ConocoPhillips, and Shell all denounced the move publicly.[33] Although how genuine they were is up for debate concerning how heavily they funded Trump and cheered for this proposal during his campaign. Somewhat perplexingly, even the North Korean government weighed in, with their foreign minister calling the move "the height of egoism and moral vacuum seeking only their own well-being even at the cost of the entire planet."[34] Any move that unites Greenpeace, ExxonMobil, Arnold Schwarzenegger, and the North Korean government in its public denouncement can only be considered ill-advised. While many specifically blame Trump for the move, it is important to realize that every other Republican Presidential candidate during the 2016 election could have just as easily done the same.

There are uncertainties when it comes to climate change. We've discussed those quite openly and in great detail during the previous chapter. These uncertainties are related to climate change impacts and magnitude, with the more dire warnings proving time and time again to be the more accurate. However, the idea that there is a real scientific debate on whether climate change is real or whether it is man-made is utterly false. There is no debate.

The debate we should be having is regarding how we are to shift our economic and political framework to incentivize the use of climate-friendly technologies on a global scale. Many great minds, which could have been put to work in coming up with more creative solutions, have been left to argue for a false position. Losing any voices in this debate is nothing less than tragic. I implore you, don't lose your voice.

# MECHANISMS, COUNTING DOWN THE DAYS & OUR ECONOMY

*If I had only one hour to save the world, I would spend fifty-five minutes defining the problem, and only five minutes finding the solution.*

*– Anonymous*

LIT by 20,000 light bulbs which, if put end-to-end, would span 40 kilometers, the Eiffel Tower illuminates the Parisian night sky. It is, without a doubt, the City of Light's most iconic monument. But on the last Saturday of March, every year, between the hours of 8:30 and 9:30 p.m. local time, something peculiar happens. The Iron Lady fades to black. So too does the Great Wall of China, Times Square in New York, Dubai's Burj Khalifa, the Acropolis in Athens, and the Sydney Opera House. Why? Because it's Earth Hour.[1]

Earth Hour is an annual event, organized by the World Wide Fund for Nature (WWF) encouraging people to switch off their lights for an hour to show their solidarity in the fight against climate change. The event is meant to encourage people to reduce their energy consumption and thus their contribution towards global warming. While a benign gesture, Earth Hour inadvertently presents us with an unintended misconception.

The burning of fossil fuels generates approximately 70 percent of the world's electricity,[2] inherently meaning that the more power we use, the more carbon we emit. However, it is essential to understand that the appliances we use don't run on fossil fuels. They run on electricity. Fossil fuels are merely the most commonly used energy source for generating electricity. While using less energy may have its benefits, it doesn't make the problem go away. The answer to the problem is switching to an energy source that doesn't emit carbon.

It is important here to acknowledge that we are, quite literally, swimming in energy. If we are to consider the amount of energy used by humans—500 quintillion (a billion billion) joules of energy a year—we don't actually use very much at all. The amount of energy the Earth receives from the sun in a year amounts to about 3,766,800 quintillion joules. Plants pluck about 3,000 quintillion joules, 0.07 percent, of that energy out of the sky. All the energy we use is equivalent to a meager 0.01 percent of the energy the Earth gets from the sun.[3]

Many believe that combatting climate change requires us to forgo the living standards we enjoy today. This isn't true. While there is a correlation between carbon emissions and prosperity, i.e., richer and more prosperous nations emit more carbon,[4] the relationship isn't as straightforward as it may seem. More affluent countries use more energy, and most of our energy sources emit carbon. If the energy source is changed to one that doesn't emit carbon, the problem just goes away.

Take Norway for example. It has an arguably higher standard of living than any other country on the planet. It has free healthcare, low unemployment, and a growing telecommunications and technology sector. Poverty is virtually nonexistent. Moreover, the vast majority, 98 percent, of Norway's electricity comes from non-fossil-fuel sources.[5] Rivers and waterfalls are the fuel powering Norwegian industry and social welfare programs. The European nation does not need fossil fuels to keep the lights on. Oddly enough, however, the country's economy is heavily reliant on fossil fuel exports, namely oil.[6]

## Mechanisms

In 2016, three countries—the United States, China and India—accounted for roughly half of all carbon emissions. The largest emitter, by far, was China, which was responsible for just under 30 percent of all emissions. Next was the United States at about 16 percent, and then India with about 6 percent. This means that the remaining 50 percent is shared between 191 countries.[7] While China's emissions are currently double those of the United States, China also has a vastly higher population, 1.4 billion compared to 325 million. Moreover, the total amount of carbon dioxide put into the atmosphere by the United States, throughout its history, far exceeds China's contributions.[8] Per-person emissions in the United States are about double those of China. Contrary to popular opinion, the United States isn't the highest emitter per person in the world. That prize goes to Qatar, which emits roughly six times the emissions per person that China does. However, Qatar only makes up 0.25 percent of current global emissions, about the same as Nigeria. Yet, Nigeria has a population of nearly 200 million as compared to Qatar's 2.6 million.[9] While this sort of information might be relevant to politicians, trying to analyze global emissions in this manner is entirely useless if we are going to look at the issue scientifically. We should, at least for the time being, ignore the idea of countries and look at the physical mechanisms that produce our emissions.

In 2010, 25 percent of all global greenhouse emissions came from burning coal, natural gas, and oil to produce electricity.[7] This is the single most significant source of greenhouse emissions. The main byproduct of burning coal, natural gas, and oil is carbon dioxide. However, carbon dioxide isn't the only greenhouse gas emitted. Small amounts of methane and nitrous oxide are also emitted. Although the quantities of methane and nitrous oxide are substantially lower than the amount of carbon dioxide, they have a much higher ability to soak up heat. Methane's global warming potential is approximately thirty times that of carbon dioxide, i.e., every ton of methane put into the atmosphere is the equivalent of thirty tons of carbon dioxide. Nitrous oxide's global warming potential is just under three hundred times that of carbon dioxide. A small fraction, less than 1 percent, of the sector's emissions come from sulfur hexafluoride (SF6). SF6 is a chemical used for insulation in equipment used for distributing electricity, such as substations. Its global warming potential is approximately twenty-four thousand times that of carbon dioxide.[10]

The majority of the world's electricity is produced by using coal and natural gas. Oil is mainly used as a transportation fuel. In the United States, coal accounts for nearly 70 percent of emissions from the sector while only generating about a third of the power. The remaining 30 percent of emissions come from burning natural gas, which also accounts for about a third of electricity production. In 2015 about 20 percent of America's electricity came from nuclear sources, and 13 percent came from renewable sources. The emissions released by these sources are negligible.[10]

Electricity is utilized in homes, factories, and offices. This means that we can attribute energy use and the resultant emissions to the end user. Doing this gives energy providers, government regulators, and researchers the ability to analyze how much energy each sector uses. While this is beneficial for analyzing energy demand, it isn't helpful when examining our carbon footprint. This is because the end user does not choose the technology and fuel used by their energy provider. Let us take the example of two homes. Both are the same size; however, one house has energy-efficient appliances while the other does not. The house with energy-efficient appliances receives its energy from a coal-fired power plant while the home with non-energy-efficient appliances receives its energy from a gas-fired plant. Which of these two houses will use less energy? Of course, it would be the house using energy-efficient appliances. But even if this is the case, powering the efficient home with coal will result in higher emissions than fueling the inefficient house with gas.

Most of us don't often think of the large-scale industrial processes needed to provide us with the commodities we use every day. Industrial emissions can be split into two categories: direct emissions and indirect emissions. As you'd expect, large-scale industry uses up a lot of energy. Indirect emissions from industry are the result of taking power from the grid. We've already accounted for those, so we're now only interested in industry's direct emissions, which account for 21 percent of the world's emissions.[7] Two-thirds come from burning fossil fuels onsite for energy. The remaining third comes from leaks of natural gas and petroleum, as well as the chemical reactions needed to produce cement, iron, steel, and various other substances.[11]

Transport accounts for 14 percent of our global greenhouse gas emissions.[7] This comes from burning fuels, in nearly all cases a refined product derived from crude oil such as gasoline or diesel, for transport on road, rail, air, and sea. Gasoline and diesel account for 95 percent of the world's transport fuel. On-road transportation, which includes hybrids, gas-guzzling SUVs, sports cars, pickup trucks, minivans, and large six-wheel trucks, account for over half of emissions from this sector. The remaining half of emissions come from other forms of transport such as airplanes, ships, and trains, as well as oil, gas, and lubricant pipelines. Most of these emissions are $CO_2$ coming from combustion processes in internal combustion engines.[12]

Not including emissions released in power plants from

electricity production (indirect emissions), residential and commercial buildings, which include homes, apartment blocks, offices, shopping malls and so on, are responsible for 6 percent of global emissions.[7] The most substantial factor, which accounts for nearly three-quarters of emissions, is the burning of gas or petroleum products for cooking and heating. The other contributor, which accounts for the majority of the remainder, is waste, specifically organic waste. When organic waste, such as food or paper, decomposes in the open, it produces carbon dioxide. This isn't a problem as the carbon dioxide emitted from food waste is part of the natural carbon cycle. However, when you put it in a landfill—an enclosed hole in the ground with no or little oxygen—decomposing food produces methane, which is thirty times as potent as carbon dioxide. A small fraction of emissions comes from water treatment and leaks of fluorinated gases used in air conditioning and refrigeration.[13]

Even though we haven't accounted for all of our emissions so far, let's start to put the pieces of this puzzle together. Twenty-five percent of emissions come from electricity generation due to burning fossil fuels to produce energy. Twenty-one percent of emissions come from the industrial sector, of which two-thirds comes from burning fossil fuels to produce energy. A significant portion of the remaining third of industrial emissions comes from fossil fuel leaks. Fourteen percent of emissions come from transportation because our vehicles burn fossil fuels used to produce the energy needed for movement. Six percent of emissions come from buildings, of which three-quarters are from burning fossil fuels to provide energy for cooking and heating. Additionally, I have failed to mention that 10 percent of all global emissions come from the extraction, refining, processing, and transportation of fossil fuels. Are you seeing the pattern here?

## The Other 24 Percent

So far we have accounted for 76 percent of our emissions. About 2 percent comes from decomposing food/organic waste; 3 percent comes from industrial sources, not accounting for processes that use fossil fuels; and 71 percent comes from fossil fuels. What about the other 24 percent?

Most people tend to neglect, or even might not be aware of, the emissions generated by agriculture and land use. Emissions from land use can seem counterintuitive. Emissions from fossil fuels are easy to understand; hydrocarbon + oxygen = energy + carbon dioxide + water. The mechanism that leads to emissions from land is very different. During the process of photosynthesis, where plants convert light into food, carbon dioxide is absorbed and oxygen is released. Some of this absorbed carbon is also taken in and stored by the soil. The amount of carbon soil can absorb is dependent on different factors such as how well the soil is managed, the type of soil, and the area's climate. The process of plants and soil absorbing carbon from the air is called biological sequestration.[14] Remember the bathtub analogy from the first chapter? Emissions are the water coming out of the faucet. Natural processes that remove carbon from the air are the sink.

Plants and soil are paramount in this regard. We call them, along with other similar natural processes, carbon sinks.

The vast majority of emissions from land use come from deforestation.[15] When a tree is cut down, it is either left to rot, or it is burned, meaning that the carbon it once stored is put back into the atmosphere. We often think that forests are mainly cleared for their wood. This isn't true. During the last fifty or so years, as the world's population exploded, large corporations realized that they could make a lot of money by destroying forests to plant 'mega crops' like corn, soy, and oil palm.[16]

You would think that big business chooses to bulldoze forests to get access to the nutrient-rich soil beneath them. Not true. Rather, they do this because the land is cheap. The quality of the soil beneath most forests is usually quite poor. The nutrients are actually in the plants and trees, which are now either rotting or burnt to a crisp. After whatever remaining nutrients left in the soil and the burnt ashes of trees are used up, the deforested plot of land is useless.[17] So they move on to another patch of land, knock the trees down, emit all the carbon stored into the air, and render the area useless. Rinse and repeat.

The Amazon rainforest is the largest forest in the world. It covers over five million square kilometers, sprawling over nine South American countries. It accounts for about a quarter of all carbon absorbed by land. We used to believe that as concentrations of carbon dioxide in the atmosphere increase, plants and trees will soak up more carbon. However, the Amazon's ability to absorb carbon from the atmosphere has dropped by a third, leading scientists to believe that it has reached a carbon "saturation point."

As emissions increased during the 1990s, trees in the Amazon did take in a lot more carbon. However, the amount absorbed had flatlined by the year 2000. Coupled with this flatlining, trees began dying a lot faster. The reason for this was that trees grew much faster during periods when they absorbed more carbon. When trees grow faster, they reach maturity faster and, thus, age and die faster. It's currently unknown if this phenomenon is limited to the Amazon or is more widespread. "Until now, the biosphere has been re-absorbing a proportion of the carbon dioxide we have released through fossil fuel burning and land use change. If that re-absorption declines as suggested here, more carbon dioxide will remain in the atmosphere, thereby accelerating climate change."[19] What we have here is a double whammy. Deforestation emits the carbon stored by trees and plant life, while at the same time the amount of carbon that trees can absorb will diminish.

Modern agricultural practices themselves are also responsible for large quantities of greenhouse gas emissions. The two main greenhouse gases emitted by farming are nitrogen dioxide and methane. Nitrogen dioxide is emitted as a result of overusing nitrogen-rich fertilizers. Nitrogen is an essential nutrient for plants. About 78 percent of the Earth's air is nitrogen. However, plants can't use that nitrogen. They rely on the Earth's natural nitrogen cycle. This is one of the main reasons why farmers give their crops nitrogen-rich fertilizer. A process called denitrification turns the nitrogen in the fertilized soil to nitrogen gas. This process isn't immediate and has a few steps. One of these steps results in the production of nitrogen dioxide. Bacteria living in the soil then convert nitrogen dioxide to nitrogen. In modern farming, as opposed to nature, there is too much nitrogen dioxide for the bacteria to convert entirely to nitrogen. The nitrogen dioxide which bacteria can't convert is emitted into the air. Inversely, the mechanism that is responsible for agricultural methane emissions is far less intricate. It is the expulsion of ruminant animals, which is just a fancy way of saying cow burps.[20]

## Counting Down The Days (A Number Soup)

Making the shift from carbon-intensive technologies to those that are carbon neutral won't happen overnight, meaning that we will have to keep emitting during this transitional period. More emissions inherently means more warming. Thus, we have to decide what we might consider an 'acceptable' amount of warming and an 'acceptable' amount of total emissions.

In the 1990s, one of the reasons that politicians struggled in creating policies aimed at tackling climate change was that there was no defined limit for what might be considered an acceptable amount of warming. A panel of German scientists understood that human civilization has existed at a time when the planet's temperature has only varied within a thin margin. From that, they reasoned that keeping global temperatures within 2°C of that margin would be our best practical option. Within two decades the majority of world governments have come to agree that any warming more than 2°C is dangerous. The scientific basis for the 2°C limit is that as temperatures increase, the impacts affecting humanity will amplify. This, however, does not mean that 2°C is safe. In fact, it still poses serious risks (see Chapter 2). Nevertheless, this necessitates that we ask the question, how much carbon dioxide can we put into the atmosphere if we are not to exceed this target?

It seems like a pretty straightforward question requiring a straightforward answer. However, there are plenty of varying numbers flying around indicating, at least on the surface, that there is no agreed carbon budget between scientists.[21] The Intergovernmental Panel on Climate Change's (IPCC) fifth scientific assessment of climate change presented a variety of numbers ranging from 790 to 1,570 gigatons of carbon (GtC).[22] A 2015 report by Christian Aid, the Green Alliance, Greenpeace, RSPB, and WWF put the figure at 880CtC.[23] In 2014, Michael Raupach of the Australian National University in Canberra published a study in the peer-reviewed journal *Nature* which set the value in the ballpark of 950GtC.[24] The "father of global warming," James Hanson, published a paper saying that we could emit another 350GtC, bringing the total number to 850GtC, and still keep warming under 1.5C.[25] Without context, it appears that we have a wide range of numbers with no agreement over how much we are allowed to emit. It's a number soup.

However, the relationship between global temperature rise and greenhouse gas emissions is straightforward. Temperature change and greenhouse gas emissions are directly proportional. So why is it that the numbers vary so much? In a word: probability. Let's look at the IPCC's values. When taking into account only the warming caused by carbon dioxide emissions, while neglecting other greenhouse gas emissions (e.g., methane) as well as natural feedbacks, the IPCC presents us with three numbers: 1570GtC, 1210GtC, and 1000GtC. These are connected to the probability of our emissions to raise the Earth's temperature by 2°C. Limiting total emissions to 1570GtC has a 33 percent probability of keeping temperatures under 2°C, while 1210GtC has a 50 percent probability and 1000GtC has a 66 percent probability.

Let's put these mind-numbing numbers into perspective. We're currently going through 11GtC a year and we've already burned through 585GtC of our budget. Our total budget to have a 66 percent chance of not crossing the 2°C threshold is 1,000GtC. This means we have 415GtC left. At our current rate, we'll go through that by 2053.

We have to remember, though, that while $CO_2$ is the primary greenhouse gas of interest, it isn't the only one. We still have others like methane, nitrous oxides, fluorinated gases, and so on. How do those impact our budget? If we're going to stick to the IPCC's estimates, to have a 33 percent chance of not hitting 2°C our budget changes from 1570GtC all the way down to 900GtC. At our current rate, we'll burn through that by 2046. To have a 50 percent chance, our original 1210GtC becomes 820GtC. At our current rate, we'll burn through that in 2039. To have a 66 percent chance, 1000GtC turns to 790GtC, giving us only until 2036, eighteen years from this book's publication.

What about the other (non-IPCC) numbers? The

880GtC put out by the Christian Aid, Green Alliance, Greenpeace, RSPB, and WWF is calculated for 50 percent probability of reaching 2°C and took into account the warming caused by carbon dioxide as well as other greenhouse gasses. There is only a 7 percent variation between their figure and the IPCC's. How about the study conducted by Michael Raupach? His number is 950GtC. It represents a calculation which only takes into account carbon dioxide and has a probability of 60 percent. This is a 5 percent variation from the IPCC's figures. What about James Hansen's claim that we can emit 850GtC and still keep warming at 1.5°C? Hansen's figure represents an entirely different scenario. In this scenario, all deforestation must end immediately, and trees are planted at an astronomical rate to soak up more of the carbon we emit, thus giving us more wiggle room in the fossil fuels emissions department.

You might be thinking that a 66 percent probability is too low. Perhaps we should be more conservative in our approach. Maybe we should take an approach that gives us 90 percent certainty. The truth is that if we wanted to take such a conservative approach, we'd blow our carbon budget within the next decade. Even the most ardent optimist has to see the impossibility of stopping all carbon emissions within less than ten years. So which one of these figures shall we take? For the sake of simplicity let's narrow our choices to the numbers put out by the IPCC. What probability shall we choose? Considering the consequences, I would like to assume that you would think that having a 66 percent certainty would be the best approach. That leaves us with two choices: to limit total emissions to

1000GtC or 790GtC. This ultimately forces us to ask the question of whether we should include non-$CO_2$ greenhouse gases in our calculations.

Let's analyze the situation. Taken at face value, non-$CO_2$ gases cause warming; thus, they should be included. However, these gases tend to stay in the atmosphere for a much shorter time than carbon dioxide, which stays in the atmosphere for millennia—forever, from a human perspective. Other greenhouse gases tend to remain in the atmosphere for much shorter periods of time. Methane, for example, has a half-life of ten years, i.e., half the methane we emit will be out of the atmosphere in ten years. It is because of this that some have argued that other gases should not be factored in. Inversely, one can say that the amounts of these gases emitted are far higher than the rate at which they leave the atmosphere so we should count them. Moreover, even if they might dissipate faster, gases such as methane produce a lot more warming per ton emitted over the span of a century. With that said, the 790GtC value is considered by most to be the most accurate, and thus most believe that it is this value which should be adhered to.

**Our Economy**

The economic ramifications of limiting total carbon emissions to 790GtC are drastic. Simply put, it will require us to reduce our greenhouse gas emissions by 5 percent every year starting from 2020 up until the end of the century. Much of the world's emissions from land use and agriculture can be reduced, or even reversed, by ending deforestation and engaging in more sustainable agricultural practices such as minimizing nitrogen-rich fertilizer use. Additionally, limiting our warming to 2°C means keeping a large portion of the world's known coal, oil, and gas reserves in the ground. Current estimates suggest that staying under 2°C means keeping 88 percent of all known coal reserves, 35 percent of oil reserves, and 52 percent of natural gas reserves in the ground (there is a caveat to this which we will tackle in Chapter 6).[26] Moreover, all Arctic drilling will have to be strictly off-limits. I cannot help but recall the words of Ahmed Zaki Yamani, Saudi Arabia's minister of oil for more than twenty years: "The Stone Age didn't end for lack of stone, and the oil age will end long before the world runs out of oil."[27]

Many countries in the world are heavily reliant on fossil fuel exports to sustain their economies. Of the twenty-eight nations whose economies are most dependent on oil, nine are Arab countries.[28] The Middle East is home to about half the world's known oil and gas reserves. By extension, this means that the Middle East also has half the world's "unburnable" oil and gas reserves. The economies of Russia, Angola, Norway (odd considering it barely uses any itself), Azerbaijan, Venezuela, Chad, Brunei, Kazakhstan, Mexico, and Nigeria are all reliant on fossil fuel exports. For such countries to maintain economic stability, their economies must diversify, reduce, and eventually end their reliance on fossil fuel exports.

Most tend to imagine that countries with "oil economies" are rich and plentiful. While the Middle East is sitting on half the world's oil and gas, the country with the largest oil reserves is Venezuela. Instead of being wealthy and stable, Venezuela is on the brink of collapse. "In Caracas, the capital, men scavenge daily in the putrid Guaire River. They pour down from the barrios, raking their hands through stinking, toxic mud in the hopes of finding the tiniest bit or metal, jewelry — anything of value — that they could sell for food."[29]

During the mid-2000s, Venezuela's then-president Hugo Chavez utilized the income from high oil revenues (due to high oil prices) to implement subsidies on food and invest in the country's healthcare and education systems. This lifted over half the country's population out of poverty. Chavez did nothing to reduce his country's dependence on oil, however, and continued to spend vast amounts of money on social welfare programs. These programs depended on oil's price remaining high. In 2014, shortly after Nicolás Maduro, Chavez's successor, took power, oil prices plummeted. All of Chavez's welfare programs disappeared. Hyperinflation led to skyrocketing food and medicine prices. This threw the majority of Venezuelans into poverty once again. The extreme corruption of the Maduro government didn't help either. Chavez rigged the country's economy to keep himself in power by maintaining economic reliance on oil so that people had to rely on him and his government, through subsidies, to support a decent standard of life. Maduro continues to rig the country's economy to keep himself in power; however, now it is at the expense of the poor, through corruption and currency manipulation.[30] Thankfully, the majority of oil-reliant countries did not suffer as drastically from the 2014 oil crash, but Venezuela's example should demonstrate that the 'good times' that fossil fuel reliance brings are temporary and, if allowed to go on without end, will result in nothing but pain and suffering. Economies that are reliant on one resource are 'fragile' while economies that are more diversified are, to steal a phrase from the Lebanese trader-turned-philosopher Nassim Taleb, 'anti-fragile.'

## On Our Terms

Considering the potentially catastrophic impacts that climate change brings with it, a shift in carbon-intensive technologies to those that are carbon neutral will be inevitable. The question is, will the transformation be on our terms or not? As it currently stands, the more destructive impacts have yet to manifest. However, every day of inaction brings us closer to them. Will the shift occur before we have hit the 2°C budget or after? Will the shift harm countries with fossil fuel economies, or will they have diversified in time? Moving as fast as we can towards our individual targets is the only conceivable way we can ensure that the shift will be on our terms.

# WALKING FORWARDS & BACKWARDS

*The word "politics" is derived from the word "poly," meaning "many," and the word "ticks," meaning "blood-sucking parasites."*
   *– Larry Hardiman.*

IT IS DIFFICULT, if not impossible, to disentangle the issue of climate change from politics. Within recent years, economics and politics have become profoundly inter-linked and economic development is deeply interlinked with activities, such as energy production, transport, and land use, that are the primary drivers of climate change. Within this chapter, I hope to present you with the most paramount political arguments surrounding climate change as well as a brief history of past political action linked to the issue.

The aim of climate-related policies should be the creation of a socioeconomic and political framework which encourages a slowdown, and ultimately a reversal, in greenhouse gas emissions. The two main large-scale political agreements regarding climate change are the Paris Climate Accords and the Kyoto Protocol. Many, including myself, believe the Kyoto Protocol to be ultimately a failure and are fearful that the Paris Climate Accords will follow in its footsteps of not getting enough done. The reason for this, as we will discuss in far greater detail later in this chapter, is that both agreements fundamentally failed to introduce mechanisms that incentivize investment in technologies that do not emit carbon, while at the same time disincentivizing greenhouse gas emitting practices.

Before we can critically analyze the successes and failings of both agreements, we have to understand the underlying political mechanisms that could allow the desired framework to form. The only conceivable way to realize this framework is to enact policies that subsidize non-carbon intensive sources, end subsidies to fossil fuels, and/or enforce limits on greenhouse gas emissions. The two primary economic/political mechanisms that are often discussed when analyzing what political action is needed to limit fossil fuel consumption are a carbon tax and a carbon cap-and-trade mechanism. Before we can get into the nitty-gritty of these mechanisms, we need to know what subsidies, negative externalities, and a Pigovian tax are.

## Economics 101

A subsidy is a sum of money given by a government to a business to keep the price of a product or service provided by the company low.[1] For example, a country's government decides that it is within the best interest of its people that essential food items such as bread be cheap. For local businesses, the cost of producing bread is growing more and more expensive, which inherently means that they will have to increase the price of their product to stay profitable. The government then steps in and pays these businesses a sum of money that should cover the increase in production cost under the condition that the price of bread stays low. Thus, the companies remain profitable despite increased production costs while the price of bread remains unchanged.

A negative externality is "a cost that is suffered by a third party as a result of an economic transaction."[2] This means that neither the buyer nor the seller of a particular product or service pays the full cost. For example, a company that produces alcoholic drinks pays for the price of manufacturing and/or distributing its product. In exchange for an alcoholic beverage, a customer pays the producer the same amount of money spent on production plus an amount on top of that which acts as the producer's profit margin. Alcohol is a product that brings with it a host of negative impacts on society, for example, the risks created by people who might drink and drive. To mitigate these risks, the state, or indeed any other third party might have to spend a sum of money that would not have had to be spent had it not been for the product, for example, increasing government spending on policing to catch drunk drivers. Neither the consumer nor the producer pays for the cost to society that alcohol brings with it.

A Pigovian tax is a mechanism which forces either the producers or consumers of products and services that have a negative impact on society to pay the cost of their consequence to the state, i.e., a tax on negative externalities.[3] For example, a country with a highly overweight population may choose to implement a Pigovian tax on food products, such as sugar, which contribute to obesity. The desired outcome of a Pigovian tax is that a price increase in products with negative externalities should act to dissuade consumers from purchasing those products. Within some cases, although far from all, the money accrued through these taxes is spent on treating and mitigating the negative impacts which these products create. For example, the funds collected from taxing sugar are spent on treating those with type-2 diabetes or introducing healthier meals at schools.

## Cheap and Abundant

The main argument for why fossil fuels are by far the most utilized source of energy is that they are cheap and abundant. There is no doubt that they are still plentiful (we keep finding more of the stuff), but are they still cheap? Fossil fuel companies currently receive just under half a trillion dollars, $492 billion, globally in government subsidies. However, this is only a small part of the picture. They receive a hell of a lot more when we count up "indirect" subsidies. Indirect subsidies are not direct sums of money given to a company to keep the price of their products down. Indirect subsidies are the untaxed negative costs associated with a product borne by society, i.e., unpaid Pigovian taxes.

In 2015, a team of economists led by David Coady, the Deputy Division Chief of the Expenditure Policy Division at the Fiscal Affairs Department (FAD) of the IMF, published a paper which provides our most accurate guide to indirect subsidies received by the fossil fuel industry.[4] The indirect subsidies given to the fossil fuel industry are not only related to its impact on climate change. They also take into account the vast majority of the other effects that result from fossil fuel use, such as air quality degradation, which killed more than 6 million people globally in 2016 alone, and groundwater contamination, which threatens fresh water supplies used for drinking and agriculture. Coady's team calculated that the total subsidies, both direct and indirect, received by the fossil fuel industry account for more than 6 percent of the global GDP. To put this in perspective, six cents of every dollar spent on planet Earth is spent trying (and often failing) to treat or mitigate the environmental and health impacts of fossil fuel use. In dollars, the total amount received by the industry in direct and indirect subsidies exceeds $5 trillion. Let me repeat that again, $5 trillion, with a *t*. Every year!

This means that the price of fossil fuels on the market has been artificially lowered from its true market cost. Of our three fossil fuel sources, oil, gas, and coal, it is coal that receives the largest amount of subsidies. When we are told that fossil fuels are our cheapest sources of energy, this is no longer true. In 2016, Lazard, a financial advisory and asset management firm, released a report which presents the unsubsidized cost of energy production from differing energy sources that year. The levelized cost of coal was found to range from $60 up to $143 per megawatt hour of energy produced. The cost of using natural gas to generate electricity using a combined cycle cost between $48 and $78 per megawatt hour, making it not only cleaner than coal but also now cheaper.

In comparison, the cost of photovoltaic solar panels (solar panels used to generate electricity as opposed to heat) ranged from $49 to $61 per megawatt hour for crystalline cells and from $46 to $56 per megawatt hour for thin film panels, both at utility scale. The cost of harnessing energy from wind ranged from $32 to $62 per megawatt hour. However, as you might have gathered, the figures presented by Lazard are in actuality for the "unsubsidized" cost of energy production, meaning that the cost of fossil fuels has been artificially lowered due to unpaid Pigovian taxes. Thus, even if we discount that the price of fossil fuels has been artificially reduced, renewables are now, for the first time in history, financially competitive with their carbon-intensive counterparts, and in many cases, renewables are now the cheaper option.

## Taxing Carbon

A carbon tax is merely a Pigovian tax on carbon, i.e., a tax on the "social cost" of emitting carbon dioxide and/or other greenhouse gases. Determining what constitutes a suitable price on carbon isn't straightforward. This is because the impacts of climate change are complex and often unpredictable in their magnitude, location, and when they will occur. However, this isn't something without precedent.

In 2008, the Canadian province of British Columbia "introduced a tax on the carbon emissions of businesses and families, cars and trucks, factories and homes across the province."[6] The province incurred a tax of $10, Canadian, on every metric ton of carbon emitted, with the rate increasing by $5 annually to bring it to $30 in 2012. The taxes themselves were "revenue neutral," meaning that the revenue generated would not be spent on subsidizing renewable energy sources, but rather the funds gained from the carbon tax were used to lower other forms of personal and corporate taxation.[7] In 2012, when the tax reached $30 per metric ton, $0.25 was added to the cost of a gallon of gasoline. The increase in fuel price pushed people to drive less and encouraged businesses to invest in greater energy efficiency. The tax gained popularity, with only 32 percent of British Columbians opposing the tax in 2015.

There is often a misapprehension that such schemes might be championed by left-wing leaning political parties and opposed by the right wing. The tax plan was introduced by the British Columbia Liberal Party, a right-wing party, and was opposed by the New Democratic Party, a left-wing party, under the slogan "Axe the Tax." Ironically, the current leftist Canadian administration based their plans for a nationwide carbon tax on the scheme implemented by the right-wing British Columbia

Liberal Party. As of the time I am writing this chapter, Canada has yet to pass a nationwide carbon tax.


## Cap & Trade

Within a carbon tax scheme, the government sets the price of carbon. Additionally, while carbon taxes aim to dissuade fossil fuel use, they do not place any limits on emissions. You can emit all you want; you just have to pay the tax. Within a cap-and-trade framework, not only is the amount of carbon allowed to be emitted predetermined, but the price of carbon is determined by market forces.[7]

A government agency, usually an environmental regulator such as the Environmental Protection Agency in the United States, places a limit on the amount of carbon and/or other greenhouse gases which are allowed to be emitted for an entire sector. Each sector is given a specific amount of carbon which they are allowed to emit, i.e., they are given the "right" to release an allotted quantity of greenhouse gas. This "right" is given to companies through government permits. This encapsulates the cap part of cap-and-trade.

The permits can be bought and sold on a market similar to the stock exchange. If a company has low carbon emissions or finds a way to reduce its emissions, it can then sell part of its allotted amount to another company which would allow the other company to emit more. This encapsulates the trade part of cap-and-trade.

While cap-and-trade might seem, on paper at least, to be more advantageous to a simple carbon tax, this isn't necessarily the case. While the government does not control the price of carbon in a cap-and-trade system, they do control the cap. A soft cap can lead to a low market price on carbon, which can negate the entire point of the mechanism. This was seen in the European Union's trading scheme, which has been criticized for having an excessively high cap leading to a low market price. Additionally, a cap-and-trade system is by design far more bureaucratic than a carbon tax, making a carbon tax much easier to implement.

Moreover, even though with carbon taxes, no limit is set for carbon emissions, if a government desires to more severely restrict emissions, as will be needed if we are not to exceed our carbon budget (see Chapter 4), they can choose to ratchet up the tax in order to dissuade carbon emitting practices. It is for these reasons that I personally favor a carbon tax over cap-and-trade. Nevertheless, I do believe that a cap-and-trade system can be astoundingly effective if it is well designed and managed.

## Funding The Alternatives

Disincentivizing the use of fossil fuels and other greenhouse gas emitting practices is only one side of the coin. There must be an active effort to encourage the use of non-carbon emitting technologies. Considering that the cost per watt-hour of power produced for renewables has become competitive with those of fossil fuels, we have to ask, why should states still actively subsidize renewables? To understand this, we have to consider the cost of two things; the value of the energy and the cost of the technology needed to harness that energy source. With fossil fuels, the cost of the energy source is the market price of the fuel used, i.e., the cost of buying oil, coal or natural gas. With renewables, the energy source has no cost. You don't need to pay the sun to shine or the wind to blow. However, the cost of building the necessary technology to harness energy from the desired source is far higher for renewables as compared to fossil fuels. The majority, virtually all, of the cost of using solar to generate electricity comes from manufacturing and installing the panels, while the cost of using coal or natural gas to generate electricity is spread throughout the life of the plant. This means that while renewables are cheaper in the long term, they are more expensive in the short term.

This means that it might be more appealing, in the short term, to invest in fossil fuels over renewables, even though there is more profit to be made from renewables in the future. Moreover, many impoverished nations do not have the necessary funds to make the transition. While access to energy is something most of us take for granted, we have to remember that millions still don't have access to energy. It is my opinion that energy is a human right. Providing these people with energy as fast as possible is a moral obligation. Their governments are unlikely to possess the financial means to provide them with power from renewables and thus will, justifiably, favor providing them with energy from fossil fuels such as coal. Providing them with energy from renewables will require the financial assistance of wealthier nations. It is important here to note that this funding must not be done as a means for rich countries to gain further leverage, control, and power over poorer nations. It must be done in the spirit of human decency and dignity.

## The Need For Global Policies

The problem of climate change is not local; it is global. It can never be divorced from its planetary implications. If one country takes measures to lower its greenhouse gas emissions, that progress can be nullified by another nation increasing theirs. What matters is not how much an individual state emits but rather the total amount emitted by everyone. It is for this reason that local pledges to "do your bit" are just not enough. International cooperation, agreements, and treaties are necessary.

From the Basel Convention, which aimed to stop the transport of hazardous waste between countries, to the Montreal Protocol, which banned the use of chemicals which deplete the ozone layer (an environmental problem often confused with climate change), international environmental agreements have had a history of success. However, climate change is an environmental problem like no other. Successful past agreements aimed to adjust and finetune technologies or behavior. The scale of technological change needed to mitigate and adapt to climate change is unparalleled.

The two international agreements of concern when it comes to climate change are the Kyoto Protocol and the Paris Climate Accords. The rest of this chapter will be dedicated to analyzing these two agreements to determine how they succeeded, how they failed, and if they implemented policies necessary to create an economic and political framework which leads to reductions in greenhouse gas emissions.

## The Road To Kyoto

In 1972, the first United Nations Conference on the Human Environment was held in Stockholm, Sweden. Attending the conference were 113 delegates and two heads of state, Olaf Palme of Sweden and Indira Gandhi of India. This meeting aimed to raise awareness about general environmental concerns, something that was of little interest to the public at the time. Twenty years later, the Earth Summit, held in Rio de Janeiro, Brazil, was attended by over 30,000 people, including representatives from 179 countries, NGOs, and members of the media.[15] The intention behind this meeting was to initiate discussions around "global environmental issues that would become central to policy implementation."[16]

It was in Rio de Janeiro that the United Nations Framework Convention on Climate Change (UNFCCC), an international treaty with the objective of stabilizing "greenhouse gas concentrations in the atmosphere at a level that would prevent dangerous anthropogenic interference with the climate system,"[10] would be negotiated and formulated. The treaty came into force on the 21st of March 1994. The treaty didn't seek to establish any concrete policies in and of itself. Instead, it provided the framework from which future protocols and agreements would determine suitable paths forward.

From 1995 onwards, signatories of the UNFCCC, annually held a Conference of the Parties (COP) to "discuss and negotiate various policy measures on climate change."[11] The first significant breakthrough was achieved in 1998, during the third COP, with the formation of the first international climate treaty called the Kyoto Protocol.

## The Kyoto Protocol

For the first time in history, nations around the planet came together and chose to take direct action to lower their greenhouse gas emissions cooperatively. The ultimate goal of the Kyoto Protocol was the "stabilization of the carbon dioxide emissions at 5% below the 1990 level."[11] It took into account the fact that wealthier developed nations have had far higher cumulative historical emissions than poorer developing countries and thus only specified reductions for developed nations, excluding developing countries such as China and India from any responsibility even though they were parties to the Protocol.

The majority of the Protocol's signatories were European countries. The non-European countries include the United States, Canada, Australia, New Zealand, and Japan. While the United States was a signatory to the Kyoto Protocol, Congress failed to ratify it, excluding the US. The official narrative given for why Congress did not ratify the agreement was that China and India should have also committed to emissions reductions. This narrative is, however, suspect due to the considerable influence of interest groups who wished to hinder the agreement (see Chapter 3). Nevertheless, the Protocol came into force in 2005 after Russia ratified it. The first, and only, commitment period lasted from 2008 until 2012.

From its inception, the Protocol contained pivotal flaws that limited its ability to affect real change. The goal of stabilizing "carbon dioxide emissions at 5% below the 1990 level" had no real scientific basis, i.e., it was not based on a specific carbon budget (see Chapter 4). Moreover, the signatories to the Protocol covered an insufficient and declining portion of global emissions. China and India were excluded from the Protocol, while the United States refused to ratify the agreement. This meant that the majority of the burden fell on the shoulders of the European Union and Japan. "The percentage of the world emissions covered by the EU and Japan declined from around 40% in the late 1990s to only about 10% by the end of the first phase of the Kyoto Protocol in 2012."[11] As the economies of the United States and especially China and India boomed, so did their emissions. The economies of the EU and Japan, as well as their populations, grew at much slower rates in comparison. The emission reductions achieved by the EU and Japan were rendered negligible due to the increased emissions of the US, China, and India.

The Protocol also failed to address "emission leakages" to other countries who had not signed the agreement. As European economies shifted from manufacturing-based economies to service-based economies, many of the goods used by Europeans were no longer manufactured locally. They were imported from abroad. While greenhouse gas emissions in Europe went down due to manufacturing fewer products, emissions in non-signatory countries went up in proportion due to them producing more goods for the EU market. For example, in the EU, methane emissions from rearing livestock went down by 25 percent between 1990 and 2012. At the same time, imports of animal products into the EU increased by 30 percent between 1992 and 2013. This means that while emissions released directly by the EU went down, the emissions resulting from the EU's economic activity simply moved abroad.

## The Road To Paris

The first phase of the Kyoto Protocol was implemented between 2008 and 2012. In 2009, negotiators at the nineteenth COP, held in Copenhagen, attempted to develop a second phase for the Kyoto Protocol, which was to kick into effect as soon as the first phase ended. They intended to extend the Protocol from only covering developed nations to covering all countries within the UNFCCC. They failed, spectacularly.

Negotiators representing developing countries noted that it was richer countries who were responsible for the majority of historical carbon emissions. From this, they reasoned that the burden of reducing carbon emissions should also fall on the shoulders of more affluent countries. Moreover, many developing countries demanded financial compensation for the damages caused by the emissions of more prosperous countries. Virtually no developed nation was prepared to shoulder such responsibilities.[12] Public pressure on policymakers was weakened due to the "Climategate" scandal (see Chapter 3).

Bad faith, a lack of willingness to take responsibility, and weak public pressure all rendered the negotiations a failure, although not a complete failure.

While the Copenhagen COP demonstrates how climate negotiations can go very wrong, this does not mean that nothing was achieved. The United States, along with China, India, South Africa, and Brazil, drafted a document called the Copenhagen Accord. The Copenhagen Accord was not a climate agreement. Instead, it was a document that members of the UNFCCC should "take note of." The only thing of true substance within this document was, for the very first time, a somewhat scientific approach to determining what could be called a "safe" level of warming. It recognized "the scientific view that the increase in global temperature should be below 2 degrees Celsius." Moreover, it was at Copenhagen that the then-US Secretary of State, Hilary Clinton, announced that developed nations would allocate $100 billion in funds to developing countries, per year, for them to combat and/or adapt to climate change.

Two years after the fiasco in Copenhagen, climate negotiators met in Durban. It was in Durban that negotiators agreed that all parties would come together to "develop a protocol, another legal instrument or an agreed outcome with legal force under the Convention applicable to all Parties." It was also agreed that this global climate agreement should be formulated "no later than 2015."[11]

## The Paris Climate Accords

"History will remember this day," said Ban Ki-moon, the secretary general of the United Nations, as negotiators, diplomats, and world leaders cheered and hugged around him. On the 12th of December 2015, in Paris, representatives from over 190 countries agreed to and signed an international agreement called the Paris Climate Accords. Although it was hailed as a great victory and a "turning point," the Accords had no teeth. The primary aim of the agreement was to hold "the increase in the global average temperature to well below 2 °C above pre-industrial levels and to pursue efforts to limit the temperature increase to 1.5 °C above pre-industrial levels."[13]

To not exceed the 2 °C limit, under the agreement, each country is to submit a plan, called Intended Nationally Determined Contributions (INDCs), which details its own individual contribution towards the effort of reducing greenhouse gas emissions. There is no legal obligation for a country to meet its INDC target emissions reductions. The entire process is completely voluntary. If a state fails to meet its target, there is no legal way to penalize it. Instead, the Accords suggest a so-called "name and shame" strategy, in which countries observe the emissions reductions of other countries and "name and shame" the countries that don't meet their targets. Given that many politicians have no problem being named and shamed, such a strategy is unlikely to put enough pressure on any country not achieving its objectives.

The idea of a global carbon budget (see Chapter 4) is

not even addressed by the agreement, meaning that the reduction targets are arbitrary and have no real scientific basis. As a result, if every country in the world met the reduction target of its own individual INDC, it still would not be enough to limit warming to 2 °C. According to the eighth edition of the United Nations Environmental Program's Emissions Gap Report, even if all targets are met, it is very likely that we would hit 3 °C by the end of the century.[14]

The Accords also recognized the disparity between rich and developing countries regarding emissions. As such, it emphasized that rich countries are to help developing countries financially in their efforts to reduce their emissions. The primary vehicle for this help was the Green Climate Fund, which as we discussed earlier pledged to provide $100 billion to developing nations per year.

**The Post-Paris Landscape**

During his campaign, Donald Trump criticized the Paris Accords, believing that the deal would increase unemployment and place restrictions on the American economy. An odd analysis considering the agreement is entirely voluntary, i.e., Trump could have stayed in the agreement and done nothing to curb emissions with absolutely no ramifications. As a result of this odd analysis, on the 1st of June 2017, Trump announced that the US would be leaving the agreement.[15] Considering that the US is currently the world's second largest emitter, and the largest historical emitter, leaving the deal does nothing but undercut the efforts of every other nation on the planet. "The Paris Agreement states that countries must wait four years before withdrawing. However, legal analysts say that Trump could shorten that process to just one year by removing the United States from a 1992 UN treaty governing global climate talks, which the president has the authority to do without congressional backing."[16] To date he has not done so, meaning that if Trump changes his mind or is not re-elected to a second term, the United States can remain in the agreement.

Moreover, as the Paris agreement failed to create the necessary economic conditions for the required technology shift, many countries have struggled and will continue to struggle, to meet their own targets with global emissions still on the rise. For example, Germany, which aims to generate 80 percent of its energy from renewables by 2050, saw its emissions increase between 2015 and 2016. According to the German Environment Agency, "That failing trajectory won't change without 'massive and rapid efforts,'"[17]

Last, but not least, is the failure of the Green Climate Fund to provide developing nations with the funding needed to mitigate and adapt to climate change. The Fund, which aims to provide developing countries with a total of $100 billion per year, has so far accrued under $3 billion. Not $3 billion annually. $3 billion total! The Fund has been politicized and has been mired in infighting and corruption. Due to a lack of transparency, many of its investments, such as supplying "$9 million in loans to a renewable energy project in rural Mongolia that observers worried would be used to power coal mining,"[18] are questionable at best.


## Walking Forwards & Backwards

The history of the international political effort regarding climate change has been that of walking forwards and backwards only to end up at the exact same spot. The Paris Climate Accords and the Kyoto Protocol provide us with many valuable lessons which we would do well to learn from. While the Kyoto Protocol did set limits on emissions from signatory developed nations, emissions leakages to countries not a party to the Protocol rendered the reductions negligible, teaching us that the effort to curb global warming is global (who would have guessed when the thing literally has global in its name).

Both agreements failed to base emission reduction targets on a global carbon budget, meaning that the objectives of both deals were arbitrary and unscientific. Moreover, the Paris deal provided no enforcement mechanism concerning countries that do not meet their pledges, thus failing to give countries the incentive needed to implement their own carbon tax or cap and trade mechanisms. While the Paris agreement did state that richer countries should help their developing counterparts, developed nations have failed to do their part, and the Green Climate Fund is a shell of what it should be.

It might be best to remind politicians what is at stake here. That the ramifications of climate change are far more dire than we realize. Should we fail to act on this matter, it will unequivocally demonstrate that the entire international process is nothing but a sham. Billions of people will face unnecessary pain and suffering due to this failure. History will remember us as the generation that had the means to act but was too greedy and petty to do so.

# 6

## IMPACT

*You cannot get through a single day without having an impact on the world around you. What you do makes a difference, and you have to decide what kind of difference you want to make.*

— Jane Goodall

IT STARTS WITH A FEVER. Then come the pus-filled rashes. On the arms, legs, and face. If the eyes became infected, blindness would soon follow. Called the "most dreadful scourge of the human species," smallpox, a flesh-eating disease caused by the variola virus, killed at least a third of those it infected. Those who survived were often left scarred and disfigured.

As far back as the 11th century, scholars and healers recognized that those who survived the illness became immune to it. The practice of rubbing powdered matter extracted from smallpox sores into shallow scratches made on the bodies of healthy individuals, a process known as variolation, was first introduced by Chinese healers of the time as a means of immunizing individuals against the disease. Variolated persons would go on to develop smallpox-like rashes but with far less severe symptoms. The process was risky, as patients could go on to contract smallpox from the procedure or find themselves infected by another disease such as syphilis. By the 18th century, the practice of variolation made its way to Europe, where smallpox was claiming the lives of 400,000 people every year.[1]

Among the thousands of children that underwent variolation in 1757 was an inquisitive eight-year-old boy named Edward Jenner. Interested in science and the natural world, Jenner would go on to study medicine, surgery, and zoology. He later moved to rural England to begin his medical practice. There he noticed something peculiar. Milkmaids who became infected with cowpox, a minor infection, became immune to smallpox. He reasoned that cowpox could be used to protect people from smallpox and purposefully transmitted from person to person as a shield against the disease.

In May 1796, Jenner met Sarah Nelms, a young milkmaid who had contracted cowpox. Using material extracted from her rashes, he tested his theory, inoculating his gardener's eight-year-old son, James Phipps. Initially, James stopped eating and developed a fever. However, the boy made a complete recovery within less than two weeks. Two months later, Jenner inoculated James with a fresh sample of smallpox. No disease developed. Young Mr. Phipps was deemed healthy and immune to the illness.

When Jenner submitted his findings to the scientific journal *Philosophical Transactions*, they were rejected and, due to the controversial nature of his experiment, he was advised not to pursue his ideas further. Unrelenting, Jenner undertook many more trials, immunizing numerous children in the process, and published his findings independently two years later. He convinced colleagues of the procedure's efficacy and even supplied them with samples. By 1802, as the vaccine made its way throughout Europe, British Parliament awarded Jenner £30,000 (equivalent to £2 million today) in recognition of his work. In a letter to Jenner, U.S. President Thomas Jefferson predicted, "Future generations will know by history only that the loathsome smallpox existed and by you has been extirpated." Jefferson was right. Today smallpox stands as the only disease that has been purposefully, completely, and utterly eradicated.[2]

Although I cannot stress enough the great depth of gratitude the world owes Edward Jenner, I also cannot emphasize enough that the eradication of smallpox could not have been achieved without careful planning and international cooperation. In 1959, more than a century and a half after Jenner's discovery, the World Health Organization (WHO), which had only formed after the end of the Second World War, passed a resolution with the sole intent of wiping smallpox off the face of the earth. A mere six years later, the WHO initiated the Intensified Smallpox Eradication Program. Many were skeptical of the WHO's ability to actualize their Program due to their lack of project experience and the large-scale international cooperation initiatives required. To make matters worse, a lack of funding for the Program led to vaccine shortages. Moreover, increased air travel meant that the disease was reintroduced in many places from which it once had been eradicated.

The WHO's initial strategy was to vaccinate as many people as possible, as fast as possible, to build "herd immunity." Due to vaccine shortages, they would soon change their approach, aiming to vaccinate those within close proximity to smallpox cases. Those who were in direct contact with a person who had contracted smallpox were isolated for two weeks and treated. In 1977, just over a decade after the initiation of the Intensified Smallpox Eradication Program, a hospital cook in Somalia call Ali Maow Maalin became the last person to contract smallpox naturally. But that wouldn't be the end of it.[3]

Professor Henry Bedson of the University of Birmingham medical school believed himself to be nearing a break-through in smallpox research. The WHO had deemed it irresponsible to keep sending Benson samples of smallpox due to the lab's poor safety standards. Unable to gather the necessary funds to improve the lab, Bedson's team continued to work with the airborne virus without airlocks or proper safety equipment. The research team kept current on their vaccinations (the smallpox vaccine needs to be re-administered every few years to be effec-tive), but Janet Parker, a medical photographer working directly above the lab, hadn't had a vaccination in over twelve years.

She thought it was a cold at first. Then came the rashes and pustules. Janet, along with her parents, were quickly quarantined. The stress of the incident caused Janet's father to suffer a heart attack, killing him. While Janet's mother survived the illness, Janet herself was not as fortunate, being the last person in history to die from smallpox. Overwhelmed by guilt, Henry Bedson stepped into his garden and, while his wife was distracted, slit his own throat.

---

I began this book by stating that we are living in what is by most accounts the most prosperous period in human history. This is in no small part due to the interlinking matrix of better technology as well as sound economic and political policy, allowing said technology to flourish. I have also stated throughout this book that the technology needed to take action to combat climate change is, for the most part, well within our grasp. Rather than a technological revolution, which for the most part has already occurred, what we truly need are policies that would allow the deployment of these technologies on a global scale.

## On Mitigation & Adaptation

Action on climate change can ultimately be split into two parts: mitigation and adaptation. Mitigation is the process of reducing and even possibly reversing our greenhouse gas emissions to limit and/or reverse global warming.

Adaptation, on the other hand, is the process of changing and improving our infrastructure to make us less vulnerable to the impacts of climate change.

Some see adaptation and mitigation as two different strategies to deal with the problem, a sort of either/or scenario. Nothing could be further from the truth. Mitigation and adaptation are two sides of the same coin. Any type of global strategy to take action on climate change wouldn't be complete without both. However, talking about suitable adaptation strategies in detail within the confines of this book would be incredibly difficult for two reasons. The first is that there are still, as we discussed in Chapter 2, a lot of uncertainties related to climate change impacts regarding their location and magnitude. These uncertainties make it virtually impossible for us to adapt preemptively. The second is that adaptation strategies are local as opposed to global, meaning that the specific geographical conditions of a particular place will play an essential role in determining which approach would be best to solve each problem.

To understand this let us imagine two coastal communities facing sea level rise. One community chooses to install sea walls to protect itself from the rising tide. The second community does not have this option. The soil on which their town or city is built is porous, meaning that even if they erect sea walls, rising sea levels would literally cause water to bubble up from underneath their feet. Such a place would have to be abandoned. If provisions aren't put in place to relocate those displaced to a more inland city or town, those displaced will have become climate refugees.

The significant uncertainties related to adaptation inherently mean that we must, for the time being, focus our attention on mitigation. For the most part, mitigation ultimately comes down to replacing technologies that emit carbon with those that don't. Although it will be far from easy, limiting our warming to 2°C, and even possibly reversing it, is well within our technological grasp.

## Fixing The Fridge

In 1995 a Mexican chemist named Mario José Molina-Pasquel Henriquez received the Nobel Prize for Chemistry due to discovering, as well as communicating, the impacts of chlorofluorocarbons (CFCs) on the ozone layer. CFCs and their proverbial cousin hydrochlorofluorocarbons (HCFCs) were chemical refrigerants once found in nearly every refrigerator and air conditioner on the planet. Henriquez found that these chemicals erode the planet's ozone layer, which protects us from the sun's ultraviolet radiation. If this protective layer were to be damaged, cases of skin cancer globally would skyrocket. In 1985, a gaping hole in the ozone, located in the Antarctic, was found. It took only two years for the international community to react. Today, CFCs and HCFCs, along with other ozone-depleting chemicals once found in aerosol cans, have been largely phased out thanks to an international agreement, developed in 1987, called the Montreal Protocol. As a result, the ozone layer is now healing.[4]

While CFCs and HCFCs have been phased out of the market, their younger nephew hydrofluorocarbons

(HFCs), which acted as the primary replacement chemical, is still going strong. While HFCs don't impact the world's ozone layer, they are a very potent greenhouse gas, over 1000 times as potent as carbon dioxide. In late 2016, negotiators from over 170 countries agreed, through an amendment to the Montreal Protocol, to begin phasing out HFCs, starting with developed nations in 2019 and developing nations between 2024 and 2028. It is important here to remember that the Montreal Protocol, unlike the Paris Climate Accords, is legally binding, with countries failing to comply facing trade sanctions. HFC replacements, such as propane and ammonium, are already on the market.

At the moment, refrigerants make up a tiny percentage of our emissions. However, as more and more countries develop, the demand for air conditioning and refrigeration will increase. An additional factor that could increase our demand for air conditioning is climate change. We can expect a higher need for air conditioning throughout the globe as heat waves and elevated temperatures become more and more common. Due to this, some estimate that the recent amendment to the Montreal Protocol, if appropriately implemented, could shave off 0.5°C of warming by the end of the century.[5]

## Bringing Back The Forest

Between 2006 and 2011, Brazil reduced its emissions more than any other country on the planet. It didn't do this by reducing fossil fuel use. Rather, it did this by substantially reducing deforestation in the Amazon. This achievement was aided significantly by an international initiative

called REDD+, which aims to foster international cooperation in regard to mitigating emissions from deforestation.
REDD+ incentivizes tropical nations to reduce deforestation by compensating them for "any resulting economic
losses." Brazil and Norway were the two strongest propo-
nents of the initiative, with the latter pledging $2.5 billion
towards the goal. Brazil, on the other hand, announced
that it would reduce deforestation by 80 percent by 2020,
in comparison to its 1996-2005 average. It would later
formalize this commitment in its own national law. Brazil
would go on to meet its 2020 target in 2010, a full decade
earlier than anticipated. Norway committed $1 billion to
Brazil as economic compensation for its efforts. It did not
seek to utilize funding as a means of gaining political
control over Brazil. The money was not a "carbon offset"
(which is when a richer country pays a poorer nation to
reduce its emissions so that the richer country is allowed
to emit more), meaning that Norway did not increase its
emissions as a result of the deal. During the time that
Brazil restricted and reduced deforestation, it also enjoyed
great economic growth and continued to export large
quantities of its local produce, such as beef and soy,
during the global recession. Moreover, 10 million Brazil-
ians were pulled out of poverty due to social welfare
initiatives such as Fome Zero (Zero Hunger) and Bolsa
Familia (Family Allowances) during that time.

Brazil's success was primarily due to good governance
and the implementation of effective policies at both the
state and federal level. It expanded the area coverage of
indigenous reserves and protected areas, which now
cover more than half of the Brazilian portion of the

Amazon rainforest. It also empowered indigenous societies by granting them direct control of about a fifth of the Brazilian Amazon. Their right to the land was legally enforced, protecting them, for the most part, from illegal encroachment. Strong logging laws were developed and implemented, which aimed to stamp out illegal logging activities and sawmills. It is important here to also appreciate the critical role that the general public played. Through initiatives such as the Zero Deforestation campaign, indigenous peoples, human rights activists, and environmental NGOs and advocates placed greater pressure on the government to act. In both 2006 and 2009, environmental advocates developed "widely publicized exposés of the role the soybean and beef industries have played in deforesting the Amazon; the resulting publicity led to commitments from those industries not to sell products raised on deforested land."[6]

Unfortunately, in 2012, the same year that deforestation in Brazil reached an all-time low, the Brazilian Supreme Court initiated sweeping changes to the laws put in place to protect the Amazon. The revisions gave amnesty to those responsible for illegal deforestation on "small properties" before 2008.[7] It also lessened the requirements on deforesters to restore the areas they had damaged. Large agribusinesses and their lobbyists argued these revisions would allow the agricultural sector to flourish by not placing much red tape around it. As a result, deforestation once again began to increase in subsequent years.[8]

## The Soil Beneath Us

As the world's population continues to increase so will our demand for food. It is projected that our consumption of wheat will increase nearly 70 percent by 2050.[9] This means that our agricultural output will have to increase substantially. This increase in production would have to occur in conjunction with more sustainable farming practices. Farmers utilize fertilizers to provide their crops with the nutrients they need, nitrogen being among the most important. As we discussed in Chapter 4, the overuse of nitrogen-rich fertilizers results in high nitrogen dioxide emissions, a greenhouse gas 300 times as potent as carbon dioxide. These emissions, for the most part, can be managed through more precise and scientific farming methods.

Annual testing of agricultural soils gives farmers a more detailed glance at what is going on beneath their feet.[10] It allows them to determine whether a plot of land is overnourished, i.e., contains more nutrients than necessary, or undernourished, i.e., does not contain enough nutrients. This, in turn, gives farmers the ability to make informed decisions on the amount of fertilizer they should apply. If a plot of land is found to be overnourished, farmers may choose to reduce the amount of fertilizer used, reducing emissions while also saving them money. If a plot is found to be undernourished, farmers can then increase fertilizer usage to cultivate higher crop yields. Another cost-cutting method that farmers can use is directly injecting fertilizer or manure into the soil. When fertilizer is sprayed, a large portion of its nitrogen content is lost to the air due to volatilization.[11] This leads to fewer nutrients for crops as well as more unnecessary emissions. Direct injection eliminates volatilization, providing more nutrients to crops while using less fertilizer and resulting in fewer emissions.[12]

There is another soil management practice that can allow farms to become carbon sinks while also solving another pertinent agricultural problem. Due to current practices, we lose approximately 23 billion tons of fertile soil every year. At this rate, the planet will be bereft of fertile soil within 150 years. Most of the blame can be put on large agro-corporations for practices such as slash-and-burn farming as well as applying excessive amounts of synthetic fertilizer. A seemingly more innocent contributor to soil degradation is the practice of tilling. Thankfully, many farmers have taken note to eliminate tilling, a method called "no-till farming." This is nothing new and is a practice that dates back 10,000 years. The popularity of tilling is the result of advanced plow designs during the 18th century, which allowed farmers the ability to plant more seeds with less effort. So what is tilling? Tilling is the practice of "turning over the first 6–10 inches of soil before planting new crops. This practice works surface crop residues, animal manure, and weeds deep into the field, blending it into the soil. It also aerates and warms the soil."[13] While this is good in the short term, it has a lot of disadvantages in the long term.

Tilling removes the top layer of vegetation covering the soil, exposing it to erosion by wind and water. Moreover, the practice kills many of the microbes that play a vital role in healthy soil biology. In the long term, these effects reduce the quality of the soil, leading first to more fertilizer use and then to more emissions before the soil is ultimately completely depleted and unfit for use. With "no-till farming" the soil is left undisturbed for the most part, resulting in a more robust soil structure that allows the soil to hold onto more water. A protective layer of vegetation covers it, making it less prone to erosion. Such soil is the ideal environment for bacteria critical to soil health. With modern instruments, such as no-till tractors, farmers are able to sow seeds more quickly and more cheaply than their conventional counterparts. Such practices are expected to save farmers approximately 50 to 80 percent on fuel costs, and 30 to 50 percent on labor costs.[14]

## The Cattle Balancing Act

Within the current paradigm, the production of beef can be an environmentally detrimental process that causes high greenhouse gas emissions as well as land degradation. Direct emissions produced by rearing livestock account for approximately 7 percent of our global emissions, with beef being the biggest single emitter of them all. As populations shift up towards the middle class, their consumption of meat generally increases. This means that producing beef in a manner that is environmentally sustainable is within our best interests.

Most cows begin life similarly. When they're babies, they drink their mother's milk. As calves, they are allowed to roam free and eat grass. Between the ages of seven to nine months, conventionally reared cows are then moved to industrial feedlots. There they are kept in confined areas with barely any space to move. They are fed a grain-based diet of corn and soy, along with a "healthy" serving of growth hormones and antibiotics, to fatten them up. Then, a few months later, they're taken to the slaughterhouse. Inversely, grass-fed cattle remain on open pastures their entire life, free to roam and eat the food they were intended to eat. Due to them not being force-fed and hopped up on growth hormones, grass-fed cows tend to live longer.[15] While grass-fed beef is far more humane than its grain fed counterpart, it is still a significant greenhouse gas contributor.

Within nature, cattle intensely graze on a particular plot of land for a period of time and then move on to another plot. This stimulates the grass in the used plot, resulting in faster growth and leading to more carbon being sequestered into the soil. In comparison, in modern cattle rearing, cows repeatedly feed on the same plot of land. This means that the grass is never given the necessary time to grow, making it unable to sequester carbon and therefore there is nothing to offset the cattle's emissions. So what if our rearing methods were done in a way that mimics nature?

Adaptive Multi-Paddock (AMP) grazing is a modern rearing methodology that aims to replicate nature's way of balancing greenhouse gas emissions. Within AMP grazing, cows graze for short periods of time in small paddocks. They are only allowed to eat 50 percent of the vegetation within a plot before they are moved to another plot. This means that, just as in nature, the grass is stimulated, which results in more vigorous grass growth. Faster grass growth, by extension, means more carbon seeping into the soil, which in turn provides the microbes in the soil with more food, who in turn provide the grass with more nutrients. These pastures don't require any fertilizer as the nutrients and water ingested by the cows ends up back on the soil as manure and urine.[16]

As this cycle is repeated, more grass grows, which results in more carbon sequestered into the soil, which in turn provides the grass with more nutrients, which allows the grass to grow faster, which then sequesters even more carbon, on and on in a self-regenerating loop. Such a loop is carbon neutral, i.e., the greenhouse gases emitted by the cattle are entirely negated by the greenhouse gases absorbed by the grass and soil and may even act as net carbon sinks, meaning that the amount of greenhouse gas taken up by the grass and soil is greater than that emitted by the cattle.[17]

## The Stuff We Throw Away

About 6 percent of our emissions come directly from buildings. Around two-thirds of these emissions come from burning fossil fuels for cooking and heating. Mitigating these emissions, on a technological level, is quite easy. Electrical heaters and cooking appliances have long been on the market. While most in the developed world rely on electric or natural gas stoves, it's estimated that somewhere between 600 and 800 million families are still reliant on solid fuels (coal, charcoal, and wood) to cook food.[18] Irrespective of global warming, burning such fuels indoors means that the pollutants cannot disperse, lingering in poorly ventilated houses. It is estimated that indoor air pollution from burning solid fuels kills 3.5 to 4 million people annually as well as causing "major health problems like COPD (chronic obstructive pulmonary disease), asthma, lung cancer, pneumonia, and respiratory tract infections."[19]

Raising awareness concerning indoor air pollution, as well as improved economic conditions are needed to convince people to swap out cooking with coal, charcoal, and wood to using cleaner alternatives. Natural gas stoves are cleaner, producing far less indoor pollution and emitting far less carbon. The same also applies to heating systems. With that said, natural gas systems come with their own set of unique risks such as gas leakages. As such, electrifying our cooking and heating systems not only mitigates greenhouse gas emissions, it also solves a variety of health and safety concerns, making it a good idea regardless of climate change.

The remaining third of direct greenhouse gas emissions from buildings comes from refrigerant use (which we've already analyzed) and waste generation. Waste production is a necessary part of life, you can't ever really get away from it. The problem when it comes to global warming isn't all waste, that is an entirely separate environmental issue, the issue is organic waste such as paper, cardboard, wood, cloth, and food scraps.

When organic waste decomposes in nature, it emits carbon dioxide. This, however, isn't a problem as the carbon dioxide emitted by organic waste is part of the natural carbon cycle. The problem arises when we send the waste to landfill. The conditions inside a landfill are anaerobic, meaning that there's no oxygen present. This means that when organic waste decomposes inside a landfill, rather than emit carbon dioxide, it emits methane, which is 30 times more potent a greenhouse gas. There are two ways for us to mitigate these emissions. However, they are not equally preferable to one another, the first being composting, which is the more preferable option, and the second being using waste to produce energy, which is the less preferred option.

Composting, to put it in the simplest of terms, is a process in which you put organic waste in one side and get nutrient-rich fertilizer out the other. In this process, bacteria feast on the waste, generating heat and carbon dioxide. Remember, the carbon emitted here is part of the natural carbon cycle, so it is not a problem.[20] Compost piles can get quite hot, reaching temperatures up to 60°C.[21] Composting is something that you can do in your own garden, or it can be done at an industrial scale.

Generating energy from waste, in its most basic form, involves incinerating waste to generate heat, which is used to produce steam, which is used to generate electricity. While waste-to-energy plants are, when considering climate change, preferable to landfills, they also pose some serious health and environmental concerns. The emissions from older plants were found to contain high amounts of lead, mercury, cadmium, nitrous oxide, sulfur dioxide, hydrogen chloride, sulfuric acid, and particulate matter.[21] While newer plants, which utilize more efficient burners along with air pollution prevention technologies such as filters and scrubbers, have substantially reduced these emissions, a small portion of these substances still ends up in the air we breathe. Additionally, unlike composting, waste-to-energy systems do not require that we segregate our waste, separating organics from inorganics. Everything gets burned. This inherently means that waste-to-energy is not carbon neutral, with the carbon locked into many non-organic products (such as plastic) being emitted into the air. Nevertheless, the total quantity of greenhouse gas emitted by a waste-to-energy plant is far less than those emitted by a landfill.[22]

## Energy

Accounting for approximately 71 percent of our emissions, burning fossil fuels to generate energy is the single most significant contributor to climate change. Up until recently, fossil fuels have been a boon to the world, providing us with cheap and plentiful energy. Up until the last few years, considering replacing them was an absolute pipedream. However, we are now witnessing something of a miracle. Something that the most ardent optimist might have never dreamed of. The price of renewables now rivals that of fossil fuels and is often even the cheaper option in the long term.

Transforming our energy infrastructure to run on renewables, rather than fossil fuels, will be nothing short of an energy revolution, on par with the rise of fossil fuels themselves. The mitigation measures we have analyzed thus far have involved tweaking and adjusting our current practices. Regarding energy, this cannot be the case. A complete overhaul of our energy systems is necessary if we are to limit warming. This change will not happen overnight, nor can it, nor does it have to. Instead, it must be brought into reality, throughout the globe, at a measured, but fast, pace throughout the following decades. Keeping our emissions below 2°C involves reducing our emissions by 5 percent every year from 2020 up until the end of the century.

As it currently stands, in 2015 humanity consumed 168,519 terawatt hours of energy.[23] The entirety of that figure was derived from our total power generating ability (the amount of energy we can produce in one hour), which currently sits at just under 18 terawatts.[24]

At their very essence, renewable energy sources are those that won't run out. These sources are:

1. Biomass
2. Wave
3. Tidal
4. Hydropower
5. Geothermal
6. Nuclear
7. Wind
8. Solar

## Baseload vs. Intermittent

The renewable revolution will not only entail a change in our energy sources, it will also entail a great deal of change throughout our entire energy infrastructure. This is because our current energy grid is inherently designed for baseload power systems while many renewables are intermittent.

A baseload system is simply a system that provides a

continuous and stable supply of energy throughout the year. It is also often the case that the amount of energy produced is under our control at all times. For example, if the demand for energy increases during a certain time of day, a coal power plant can simply burn more coal to meet this demand. This is of paramount importance. In virtually every electrical grid on the planet, energy is consumed at the exact same instant at which it is generated. A variety of problems may arise if, at any moment, the amount of energy generated is not equal to the amount of energy consumed.

An intermittent source is one that sometimes generates power and sometimes does not. For example, on a windy day, a wind turbine may generate an enormous amount of energy while not generating any power on a day with no wind. With our current energy infrastructure, this can be problematic. For example, if, on a windy day, a wind farm produces too much power, it might overload the electrical grid. On the other hand, if there are periods during which the wind does not blow, the amount of energy generated might not be enough to meet our energy demand.[25] Just as with renewables themselves, the technology needed to deal with intermittent energy generation is well within our grasp. In fact, I am willing to bet that you use that technology multiple times a day.

## Biomass

Plants utilize photosynthesis to soak up energy from the sun and store it as carbohydrate. Fossil fuels are in essence plant matter concentrated over millions of years.

Just like coal, oil, and gas, plant matter can be burned to generate energy. Biomass is the technical term we use to refer to plant matter utilized in this regard.[26]

Biomass is considered to be a renewable energy source. This is because the plants used to produce biomass energy can be replaced within a short period of time, especially when we consider that coal, oil, and gas take hundreds of millions of years to form. Many things set biomass apart from other renewable sources, the most important being it is the only one that isn't necessarily carbon neutral. The chemical reaction at the heart of biomass is actually exactly the same as for fossil fuels: hydrocarbon + oxygen = carbon dioxide + water. However, the carbon cycle dynamics at play with biomass are very different from that of fossil fuels. With fossil fuels, carbon stored in the ground for millennia, i.e., has been out of the natural carbon cycle for millions of years, is released into the atmosphere. Biomass, on the other hand, toys with the carbon already in the natural cycle. When plants die and decompose, they emit carbon dioxide. Plant life, when left on its own, regenerates and reproduces at a proportional rate, resulting in the same amount of carbon emitted by dying plants being sequestered by newer plants. This means that biomass can only be considered carbon neutral if the plant life used for energy is replaced at the same rate at which it is cut down. The result is a self-regulating cycle of emissions being offset by new plant growth.[21]

Biomass can seem like an attractive option because it is a baseload energy source, meaning that it can be easily integrated into our current energy infrastructure on a large scale. Additionally, many coal- and gas-powered plants can be retrofitted to utilize biofuels (fuels derived from biomass that can also be used as a replacement for gasoline or diesel in cars).

Some believe using more biomass might buy us some much needed time. This is because other, more intermittent, renewable sources will require additional adjustments to our energy infrastructure if they are to be deployed at the necessary scale.[21] However, separate studies conducted by the United States EPA, Energy Information Administration, and the Union of Concerned Scientists all indicate little economic potential for biomass considering its high costs compared to other renewables.[27] Considering the relatively high costs, as well as the balancing act required to keep it carbon neutral, I cannot envisage biomass to be anything more than a supplementary form of energy for most countries.

## Wave & Tidal

Waves form due to the wind blowing on the surface of the ocean. These are strongest in open waters as there is no land mass in the way of the wind. Utilizing wave energy isn't a new phenomenon, researchers have been analyzing its potential since the 1970s. Today the technology has evolved to a point at which commercial-scale demonstrations have been deployed. At present, over a hundred wave energy projects are being developed.[28]

Both fossil fuels and biomass generate energy in a similar fashion. They are burned to produce heat, which produces steam, which turns a turbine connected to a generator, which produces electricity. Oddly enough, wave energy converters utilize a similar principle. Oscillating Water Columns, the most widely used and simple form of the technology, uses a semi-submerged chamber containing a pocket of trapped air. As waves move up and down, they force air in and out of the chamber. This movement of air turns a rotor connected to a generator, which produces electricity.

While waves and tides are similar, they are two distinct and separate physical phenomena. Although they both involve utilizing the energy of the ocean, waves and tides have different capacities concerning their ability to generate power.

Waves are formed due to the wind blowing on the surface of the water, tides, on the other hand, form due to the gravitational pull of the moon. As the moon moves in its orbit, its gravitational pull creates a tide which itself circles the globe. The movement of the Earth in space causes another tide to form, traveling in an opposite direction to the one created by the moon. These two waves are the two high tides, between which lower tides occur. The difference between high and low tides, which can reach 12m, holds an immense amount of potential energy. At the same time, the movement of water due to tides carries with it large quantities of kinetic energy. We can tap into both of these energy forms to generate power.[29]

While tidal is considered to be intermittent, it is also considered to be predictable. This is because, unlike wave energy, tidal energy isn't influenced by the weather, but rather by the gravitational pull of the sun and moon, resulting in predictable cycles.

Both tidal and wave power simply cannot physically meet anything but a minute fraction of our energy demand. Only 2 percent of the world's 800,000 kilometers worth of coastline acts as a viable candidate (i.e., exceeds a power density of 30 kilowatts per meter) for wave power. This means that the planet's entire wave energy potential is 500 gigawatts, which is equal to less than 3 percent of our current energy capacity (18 Terawatts). While the global energy potential for tidal is double that of wave, sitting at around 1 Terawatt, it still provides less than 6 percent of our capacity. Additionally, the cost of wave and tidal is higher than every other energy source, making them, from both a financial and an engineering perspective, an unwise investment.

## Geothermal

Geothermal energy is, in essence, heat energy given by the earth, originating from the initial hot formation of the planet. Pipes are used to transport steam and/or hot water from the depths of the planet's crust, which (and you might be noticing a pattern here) turns a rotor connected to a generator, which produces electricity. It is considered to be reliable and cost-effective (however this is heavily dependent on the location of a facility) while also being carbon neutral.

At 12.7gigawatts, geothermal accounted for less than 0.1 percent of the world's electricity generation capacity in 2016. Many believe that geothermal, as an established and reliable energy source, has a lot of room to grow. This is true, but not at a scale that will come even close to making a dent in our emissions. It's estimated that the world's power generation potential for geothermal sits at around 200 gigawatts. However, this potential can only be realized with future emerging technologies that might allow us to tap into previously inaccessible areas. Considering that if the full potential of geothermal is accessed, it would be equal to just over 1 percent of our energy capacity, it would be more useful for most countries to invest their time and effort on other renewables.[30] This, however, is not a blanket statement. In countries such as El Salvador[31] and the Philippines,[32] where geothermal already accounts for a quarter of their electricity generation capacity, researching ways to make better use of this resource could be economically beneficial.

## Nuclear

There is some debate on whether nuclear should be considered a renewable form of energy generation. Regardless of this debate, I have chosen to include it in this analysis as it is a non-carbon intensive energy generation process.

Nuclear fission is by far the most complicated method humans have concocted to generate electricity. However, at their heart, nuclear power plants are incredibly similar to all the previous energy generation systems we've analyzed so far. To simplify it down substantially, nuclear fission is when an atom is split in two. When this happens, an enormous amount of energy, in the form of heat, is released. This energy is then used (you guessed it) to heat water, which generates steam, which is then used to turn a turbine connected to a generator, which produces electricity.

Nuclear is, all at once, incredibly safe and stupendously dangerous. The radiation emitted by nuclear energy can damage our cells and DNA, which can result in cancer as well as hereditary mutations. If that radiation is intense enough, it could kill you instantly. However, nuclear power plants don't release much of that radiation into the environment. Whatever is released comes from discharges of water and steam used to turn the turbine generator. The rate of radiation release is always regulated by an external regulatory body, for example, the Nuclear Regulatory Commission in the United States. Under current practices, the rate at which power plants emit radiation into the environment may increase a person's, living downwind of the plant, risk of cancer by 0.1 percent.[32] However, when most people talk about the dangers of nuclear energy, they are not referring to the hazards of routine plant operation. They are relating to low probability high-risk incidents such as the nuclear meltdowns that happened in Chernobyl, Fukushima, and Three Mile Island.

Of the three, Chernobyl was by far the worst. While only 30 people were killed as a direct result of the incident, the World Health Organization estimates that the actual number of deaths from Chernobyl, due to cancer resulting from radiation exposure, to be at least 4,000 with higher estimates putting the figure closer to 9,000 deaths.[34] The Fukushima incident, on the other hand, resulted in no direct casualties. "A 2017 report from the United Nations Scientific Committee on the Effects of

Atomic Radiation concluded that health effects to the general public from radiation were almost nil. The committee expects to see two or three more cancerous tumors among the 173 workers most exposed to radiation."[35] However, the quick evacuation of over 110,000 people resulted in the deaths of 1,600 people, leading analysts to conclude that, despite the radiation, it would have been safer for everyone to stay where they were. Three Mile Island, which unlike Chernobyl and Fukushima was only a partial meltdown, resulted in no casualties. The small amount of radiation released into the environment was found by the Nuclear Regulatory Commission to have "no detectable health effects on plant workers or the public."

A nuclear meltdown happens when the portion of a nuclear reactor containing the fuel rods isn't cooled properly. Older reactors used pumps, requiring electricity, to move cooling water through the system. If the pumps fail, such as in Fukushima, you're screwed. But newer plants don't use pumps. Instead, they rely on gravity to move cooling water from elevated tanks to the core, substantially reducing the risks posed by nuclear power plants.[33]

Many see the nearly 400,000 tons of waste produced by nuclear power plants as a significant reason not to invest in nuclear. While nuclear waste is indeed a serious problem, it is essential to look at it within its proper context. Coal power plants in South Africa alone produce over a hundred times the amount of waste generated by every nuclear plant in history in one year. Moreover, nuclear waste, which is made of metal rods containing spent uranium, is for the most part well managed. The rods are initially placed in pools of cooling water. After the rods have been cooled enough, which can take a few years, they are placed in helium filled metal and concrete containers. These containers are then monitored to ensure that nothing leaks into the environment. Since 1986, when the process of dry storage first began, the process "has released no radiation that affected the public or contaminated the environment."[35] The majority of legacy nuclear waste predicaments are related to quantities generated before this date.

The stringent safety precautions taken by the industry are expensive, costing providers billions of dollars and thousands of man-hours. It is economics, rather than engineering, which holds back nuclear from being a more widely adopted energy source. As it stands presently, nuclear energy is only behind wave and tidal as the most expensive way to generate electricity. Many nuclear plants never turn a profit. However, Generation 4 reactors, which are still in their research phase, aim to standardize design, a practice common in the fossil fuel industry, to increase safety further, bring down construction time, and substantially reduce cost.[22] Whether the potential of Generation 4 reactors will be realized within the necessary timeframe is unknown.

Another impediment to nuclear is security. Nuclear stands alone as the only weaponizable energy source. The worst thing you can do with a solar panel is throw it at someone. Uranium 235, the type of uranium used in both nuclear reactors and nuclear bombs, makes up only 1 percent of the uranium naturally found on earth. Uranium 238 makes up the other 99 percent. "The concentration of uranium 235 needs to be increased to about 5 percent (low-enriched uranium) for nuclear reactor fuel and to about 90 percent (highly enriched uranium) for nuclear bombs."[36] To achieve such concentrations, uranium has to be enriched using high-speed centrifuges. These machines, found in every nuclear facility in the world, can be used to produce both energy grade uranium and weapons-grade uranium, meaning that any "civilian enrichment facility can be used to produce nuclear weapons material."[36] Because of this, all facilities owned and run in, or by, a non-nuclear state is regulated by the International Atomic Energy Agency (IAEA), who ensure that the facilities aren't secretly being used to produce weapons.

It is the mix of security risks and the lack of economic incentives which leads me to believe that nuclear won't expand much as an alternative energy source. Generation 4 reactors may surprise us by bringing down costs substantially; however, until they physically materialize, I cannot envision the nuclear industry changing much from its current state.

## Hydropower

In a 1905 article, the inventor Nicola Tesla wrote, "Coal and oil must cease to be important factors in the sustenance of human life on this planet. It should be borne in mind that electrical energy obtained by harnessing a waterfall is probably fifty times more effective than fuel energy."[37]

Hydropower is a well-established and mature renewable energy source that's been around since the late 19th century. It is without a doubt the reigning renewable king, making up about 16 percent of the planet's current electricity generation capacity. In addition to generating electricity, hydroelectric dams "also provide other key services, such as flood control, irrigation and potable water reservoirs."[37] So far, we've only tapped into about a quarter of the world's hydroelectric capacity, meaning that while hydropower can never be the only energy source we rely on, it can play a significant role in our journey towards transitioning away from fossil fuels.

Hydroelectric dams utilize the movement of water from a river to (say it with me) turn a turbine connected to a generator, which generates electricity. It's considered to be a baseload energy source. Whenever demand drops, the amount of water flowing through the turbines can be reduced, and whenever demand increases, more water can be made to flow through the turbines.

Hydroelectric dams are major engineering projects, while their costs are often location specific they can go up to the tens of billions of dollars. Moreover, hydroelectric dams can have significant environmental ramifications, impacting fish and bird migration patterns, degrading water quality, and eroding river basins. To get around these hurdles, we have in-stream hydropower generators, which in essence are high-tech water wheels.[22] While the amount of energy generated by in-stream generators is far less than that produced by a hydroelectric dam, they are far cheaper to build and pose a substantially reduced risk to the environment. This often makes them viable options where hydroelectric dams would not be feasible.

## Wind

Wind turbines utilize the movement of the air to turn a turbine connected to a generator, which generates electricity. Currently, wind makes up 3 percent of the world's energy generation capacity, but due to their commercial appeal wind turbines will make up a substantially higher proportion of the globe's generation capacity in the coming years. This is because the cost of wind has dropped significantly within the past three decades, making it the cheapest energy source on the market. The levelized cost of coal ranges from $60 to $143 per megawatt, while the levelized cost of wind ranges from $32 to $62 per megawatt.[38] This drop in price has made it an incredibly appealing, if not the most attractive, energy source for financial investors to put their money in.

The world's energy potential for wind, to put it mildly, is astoundingly enormous. Our current energy generation capacity sits at 18 terawatts. On-shore wind alone has the potential to provide 95 terawatts, 500 percent more than all of our energy needs. If the world's off-shore potential is taken into account, the world's energy potential from wind is virtually infinite. The ability of wind turbines to harness this energy has improved remarkably within the past thirty years. In 1985, the average wind turbine was rated at 0.05 megawatts. By 2014, the average wind turbine was rated at 2 megawatts, 4,000 percent higher than its counterpart twenty years before. The largest on-shore turbines are rated at 8 megawatts, more than 16,000 percent higher than those produced in 1985. While the span of turbine blades has increased remarkably, often spanning more than 100 meters, the actual turbines themselves don't take up much land. This has led to partnerships between wind energy providers and farmers. The wind generators lease a small portion of a farmer's land, which saves them money while also putting more money into the pockets of farmers.[39]

Moreover, the time needed to put up wind turbines on an industrial scale is short. Nuclear power plants and hydroelectric dam projects are massive engineering undertakings that take years, if not decades, to begin operation. Wind turbines, on the other hand, can be put up quite quickly. Within a mere seven years, Germany nearly doubled the amount of energy it gets from wind, from 9.3 percent in 2010 to 18.7 percent in 2017.[40, 41] Due to the dropping price of wind power, many companies are now bidding to construct off-shore wind farms without any governmental subsidies, simply because it makes financial and economic sense. The rise of wind in Germany has proven to be immensely popular with the general public. Hundreds of thousands of ordinary citizens have invested their money into "civilian" wind farms. Thousands of smaller companies, in essence, small-scale energy providers, have blossomed within the last decade. This is nothing less than revolutionary considering that energy companies have historically always been multibillion-dollar corporate monopolies. The wind energy sector employed more than 140,000 people in 2015.[42] In comparison, the German coal industry currently only employs 30,000 people.[43]

All of this doesn't mean that wind is perfect. It has some serious issues that must be addressed within the coming years if it is to continue to make its meteoric rise. Wind, by its very nature, is intermittent. Unlike tidal energy, its intermittency isn't predictable. Days, or entire weeks, can pass without any wind. Additionally, wind can come in intense bursts which generate immense amounts of energy in short periods of time. For example, in Germany, in 2017, extreme winds caused an astounding amount of energy to be generated, the equivalent of 40 nuclear power plants going online instantly. Grid regulators struggled to ensure that the nation's grid wasn't overwhelmed. They actually paid people to use electricity to ensure that a grid overload wouldn't occur. This poses a severe risk to our energy systems. This is because, within our current systems, energy has to be generated at the very instant at which it is used.

If we are to rely more on wind energy, we have to invest in grid energy storage. Without energy storage, powering a country on wind alone just isn't possible. Using our current practices with wind is the equivalent of feeding a dog broccoli and then complaining that there's something wrong with the dog for not eating it. The engineering problem isn't the wind turbines. It's the grid.

It is important here to note that the technology needed to update and upgrade our electrical networks exists and will be covered in a later section. If we are to look at wind alone, without taking into account issues of energy storage, it is indeed a cheap and infinite energy source that will play a paramount role in our energy future. However, there is one other energy source that has far more potential and utility than wind.

## Solar

On average the sun releases 384 yokka watts of power in the form of light and heat (that's 38,400,000,000,000,000,000,000,000 watts).[44] Only a small fraction reaches the surface of the earth, about 174,000 terawatts (174,000,000,000,000,000 watts).[45] To put this into perspective, with an 18 terawatt capacity, we used a total of 168,519 terawatt hours of energy in 2015, meaning that the amount of energy the earth receives from the sun in a single hour is more than the amount of energy used by humanity in a single year.

When we stop to think about it, all the energy we use really comes from the sun. Biomass is nothing more than solar energy captured by plants and stored as carbohydrate. Oil, gas, and coal are essentially biomass concentrated over millions of years. The hydrologic cycle, which provides us with hydroelectric power, begins with the sun heating up the surface of the oceans. This causes water to evaporate. Warm air carries this water to higher elevations, where temperatures are cooler and clouds can form. The water then comes back to the surface of the earth in the form of rain, feeding the rivers we build hydroelectric dams on. Wind is caused by the uneven heating of the earth's surface by the sun. Waves are the result of wind blowing on the surface of the water and thus are also ultimately caused by the sun. So, if all the energy we use really comes from the sun, why not cut out the middleman?

In 1873, an electrical engineer named Willoughby Smith found that electrical conductivity of a chemical called selenium, often found in metal sulfide ores, increased when it was exposed to sunlight. Three years later, Professor William Grylls Adams of Kings College along with his student Richard Evans Day determined that selenium could be used to produce electricity from sunlight. In 1883, only a decade after Smith's original discovery, an inventor named Charles Fritts built the world's first solar panel using selenium wafers. However, the solar panels of today aren't made from selenium, they're made from silicon. The invention of the silicon-based photovoltaic solar cells of today can be attributed to the work of Daryl Chapin, Calvin Fuller, and Gerald Pearson at Bell Labs in 1954.[46]

The efficiency of the initial Bell Labs solar cells was a mere 4 percent. Comparatively, a coal-fired power plant is approximately 35 percent efficient. However, within mere months the efficiency of solar cells increased to 6 percent and by 1960 scientists at Hoffman Electronics figured out how to increase the efficiency of solar panels to 14 percent.[47] Nevertheless, utilizing solar energy was far too prohibitively expensive to be even put into consideration as a utility-scale energy source, costing $250 per watt.[48] Due to their reliability, they were only used to generate energy for satellites in space. Within less than two decades, by 1977 the cost of solar dropped to a third of what it once was, costing $76 per watt.[49] As the years went by, the efficiency of solar continued to increase while its price dropped lower and lower. By the mid-1980s, researchers at the University of South Wales developed a solar panel with a 20 percent efficiency. By 2009, the levelized average cost of utility-scale solar had fallen to \$350 per megawatt (one million watts). In 2016, researchers at the University of South Wales went on to break their own record by developing a solar panel with an efficiency of 35 percent. However, most panels on the market still have efficiencies between 15 and 20 percent (it takes time for cutting-edge technology to hit the market). Since 2009, the cost of solar has dropped by 86 percent to \$50 per megawatt,[50] making it half as expensive as coal and putting it neck in neck with natural gas. Drops in solar's costs have consistently outpaced predictions, and it is not controversial to say that solar will beat wind within the coming few years to be our cheapest source of energy. Being cheap, infinite, and reliable, I would not be alone in predicting that solar will beat out every other energy source in the long term to become the world's primary source of energy before the end of the century.

Many see solar's intermittency as its Achilles heel. This just isn't true. While solar is intermittent, it is incredibly predictable. I can comfortably say that the sun will come up tomorrow. If it doesn't, transitioning away from fossil fuels would be the least of our worries. The amount of energy and the number of sunlight hours a place will receive every day is easy to calculate. In fact, it's something we already do as part of daily weather reports. While not perfect, our daily weather predictions tend not to be far off from reality. The goal with photovoltaic solar isn't the same as that of coal, gas, oil, biomass or nuclear, i.e., generating energy at the same rate at which it is used. Instead, the goal with photovoltaics is to produce all the energy a town or city will need for an entire day during sunlight hours and storing that energy in a way that's easy to access when it's dark.

While photovoltaic panels are the most famous and widely used method to harness the sun's energy into electricity, there is another method that has immense potential and negates much of solar's dependence for success on updating the grid. Concentrated photothermal systems coupled with molten salt reactors are able to store energy and provide it on demand. Unlike photovoltaics, which utilize silicon-based panels to directly convert the energy from sunlight into electricity and have no moving parts, photothermal molten salt reactor systems employ a vast array of mirrors to concentrate heat from the sun on a particular spot, in this case a tank filled with salt placed on top of a high tower (often nearly 200 meters high) in the middle of the array. The salt can reach temperatures over 500°C and can retain this heat for a great many hours. The heat, stored inside the salt, is then used to boil water, producing high-pressure steam, which is used to (you guessed it) turn a turbine connected to a generator, which produces electricity.[21]

Photovoltaics remain more cost competitive than photothermal systems. However, the cost of photovoltaics only encompasses the cost of generating electricity while photothermal systems both produce and store energy. Moreover, the cost of utilizing photothermal systems currently sits at about $100 per megawatt making it cost competitive with coal.

One advantage that photovoltaics will always have over photothermal systems is their flexibility. It's not uncommon in many parts of the world to walk by a house with photovoltaic solar panels on the roof. They allow virtually anyone, in an area that has applied net metering, to produce their own energy and sell it back to the grid. While utilizing large plots of land for solar farms, which in essence act in the same way as power plants, will be necessary, photovoltaics, unlike any other form of energy generation, can be incredibly local and allow people to build their own energy infrastructure.

Solar, as well as wind, is dilute when compared to coal, oil, gas, and biomass. The reason for this is that fossil fuels are, in essence, highly concentrated bundles of solar energy. This inherently means that the land needed to generate energy from solar and wind greatly exceeds the amount of land required for coal and gas. The amount of land needed to power the entire planet on solar is approximately 500,000 square kilometers, an area roughly the size of Spain.[51] At first glance, this can seem like an astonishingly large amount of land. It is. But, as with all things, we have to look at this figure in its proper context.

The planet has about 148,326,000 square kilometers of land. While human activity impacts the vast majority of the biosphere, we only occupy 10 percent of the Earth's landmass, which for the sake of simplicity we'll say is 15,000,000 square kilometers.[52] This means that powering the planet entirely on solar requires us to use just over 3 percent of land currently inhabited.

If we were to fill the entire Sahara desert, an uninhabited wasteland covering 9 million square kilometers, with solar panels, this would provide us with 630 terawatts of energy capacity, over 3,700 percent of what we currently use.[52] "The UAE has plans to construct 1,500MW of capacity by 2020 which will require a space of 3 km per side. If the UAE constructed the other 7 km per side of that area, it would be able to power itself as a nation completely with solar energy."[53] If the United States were to go completely solar, it would require an area equivalent to 26,000 square kilometers.[54] If we were to compare this to the amount of land used by coal-fired power plants, there's no question that coal would be seen as requiring far less land. However, this simple assessment doesn't take into consideration the amount of land needed to mine coal. The area of land in the United States currently dedicated to mining coal is estimated to be over 77000 square kilometers,[55] equivalent to three times the amount of land needed to power the US on solar. Moreover, the area of land lost to oil and gas exploration in the United States, since 2009, is equal to 30,000 square kilometers,[56] greater than the amount of land needed to power the entire country on solar.

## The Lithium Grid

Electrical grids are incredibly complex systems. At every moment, the supply of electricity must be equal to the amount of energy used. As such, grid regulators and power plant operators work in tandem to ensure that the necessary adjustments to the amount of energy generated are in place to ensure that it is equal to demand, which varies constantly. If too much electricity is put into the grid, roaring blackouts will soon follow. On the opposite side of the spectrum, an excess in demand and a lack of supply is also problematic. In Texas, during a particularly cold morning in March 2014, the demand for electricity skyrocketed as people turned up their warmers full blast. During that period of the year, due to low average demand, many power plants in Texas go offline for maintenance. Due to high demand and low supply, on that fateful day the cost of electricity ballooned to $5,000 per megawatt, a hundred times its average cost.[57]

When we consider that the two renewable energy sources with the greatest potential, solar and wind, are both intermittent, we have to face facts and accept that our current inflexible grid systems just aren't up to the task. The grid has to become more flexible. Additionally, when we consider what moving beyond fossil fuels entails, it means electrifying many systems (such as transport) that currently utilize combustion. This means that electrical systems will have to be more reliable and resilient than they are today.

If we stop to think about it, the way we use electricity is quite bizarre. From food in fridges to oil in tanks, we store every other resource we value. Electricity is the odd one out. Storing electricity at grid scale is perhaps the single most important thing we can do to reform the grid, increase its reliability and resilience, and prep it for the mass adoption of renewable energy. Rather than use energy at the very instant that we generate it, we must, as with every other resource we value, put some of that energy aside for later use.

Grid-scale energy storage isn't something that most electrical systems have in place. However, the practice of pumping water behind hydroelectric dams to provide power during times of peak energy demand is something that has been around for decades.[58] This constitutes the vast majority of the world's current grid-scale energy storage capacity. However, such systems just aren't feasible for every part of the world. Another form of energy storage that gained attention during the late 2000s was to utilize excess energy supply to split water into hydrogen and oxygen (a process called electrolysis). The hydrogen, which when burned only emits water, could then be used to generate electricity.[59] While many have dreamed of a hydrogen economy, the idea hasn't gained much traction. A form of energy storage that has gained some attention during recent years is compressed air energy storage. It is, however, still as a technology within its infancy.[60] Perhaps rather than looking towards such forms of storage to fix the grid, we should look towards something that has proved its reliability repeatedly.

Batteries, specifically lithium batteries, are perhaps the form of energy storage we are all most familiar with. Lithium batteries power our smartphones, laptops, and electric vehicles. They are so reliable that we even put them inside of people. Pacemakers, as well as many other implantable medical devices, use lithium batteries that can last fifteen years before they need to be replaced.[61] Within recent years, lithium batteries, more specifically lithium-ion batteries, have become the undisputed king of grid-integrated energy storage, making up more than 90 percent of newly introduced storage capacity.[62] For good reason.

Lithium-ion batteries are incredibly efficient, retaining virtually all of the energy put into them. They are also able to dispatch energy within milliseconds, making them incredibly flexible. This is of paramount importance as it means that lithium-ion batteries are ideal for times of peak energy demand.

In many countries, so-called "peaker plants" are used at times of peak demand to ensure that supply matches demand. These plants are often the oldest, most unreliable, and dirtiest in the fleet. They often take the form of plants that are far too old and expensive for regular use. Not only are they high carbon emitters, but they also emit high quantities of particulate matter and carbon monoxide which are detrimental to ambient air quality (the stuff we breathe). Rather than being decommissioned, these plants are retained, only to be switched on during times of peak demand. Batteries, with their ability to dispatch energy in an instant, offer us the ability to finally retire peaker plants, giving us a cleaner and more reliable solution to the problem of peak electrical demand.[63]

In the 1990s, lithium storage cost approximately $10,000 per kilowatt hour. Just as with solar, the cost of lithium-ion batteries is dropping faster than analysts have anticipated. In fact, the cost of lithium storage is falling more quickly than the price of solar. Between 2012 and 2017 the cost of lithium batteries dropped by 70 percent. In 2017, analysts predicted that the price of lithium energy storage would drop below $200 per kilowatt hour by 2019, making it much more economically feasible to apply it at grid scale. We crossed this threshold in 2018. In 2018, projections place the cost of lithium storage to be $100 per kilowatt hour by 2019 or 2020.[64] We may well have reached that threshold by the time that many of you are reading this book.

## On Hydrogen & The Future Of Transport

For millennia humanity relied on beasts of burden (horses, donkeys, and camels) to move from place to place. As we sit angrily in traffic and fall asleep on airplanes, quite literally metal tubes barreling through the sky, most of us never stop to think about the convenience and comfort that modern transportation has afforded us. We are now able to travel halfway across the world within less than a day and spend half that time grumbling. For our ancestors, such a journey would have taken years, not to mention the risk to life and limb along the way. With that being said, transport accounts for 14 percent of our greenhouse gas emissions, with cars making up about half of those emissions.

During the late 1800s and early 1900s, electric and gasoline-fueled cars vied for market dominance. During that period, many believed that the future was electric. This was because gasoline cars, while promising, posed a variety of problems. Changing gears was a hassle, they had to be started with a hand crank, they were noisy, and they stank. Electric cars, on the other hand, were quiet, a pleasure to drive, and didn't smell. Then came Henry Ford's Model T, costing less than half of what an electric car would cost, and the electric starter, which eliminated the need for a hand crank. The electric car then faded from public consciousness, only to come back into consciousness nearly a century later with the rise of Tesla Motors.[65]

Although they make up a small portion, 1.3 percent in 2017, of the global car fleet, electric vehicles have proven to not only be more reliable, due to their more simplistic design, than their gasoline-powered counterparts, but they're also cheaper in the long run. While electric vehicles are still more expensive when only looking at the cost of purchase, they require far less maintenance and use a far cheaper fuel source (electricity). A large portion of the cost of producing an electric car is in the batteries. As the price of lithium-ion batteries drops year upon year, we can expect the cost of electric vehicles to decline in tandem, meaning that electric cars within the next two or three decades will be cheaper to buy than gasoline cars.

While the path towards fossil-fuel-free cars seems relatively straightforward, at least from a purely technological perspective, the path towards carbon neutral airplanes is far less direct. Theoretically, electric planes should be quieter, more efficient, and produce no greenhouse gases. However, flying is an incredibly energy intensive process, and the current generation of lithium-ion batteries is simply not suitable for use in aircrafts. This is because they do not offer the needed energy-to-weight ratio to be ideal for use in aviation. Jet fuel is incredibly energy dense, holding 17 times more energy per kilogram than the current generation of batteries. Simply put, electrifying airplanes with batteries just isn't feasible due to the batteries being too heavy.[66]

Within the past few years, there has been a push to electrify shipping vessels. Oddly enough, the world's first electric shipping vessel, launched in China in late 2017, was used to haul coal over a short distance, 50 miles.[67] It had a maximum speed of eight miles per hour. Less than a year later, construction began on the world's first long(ish) distance electrical barges. The five barges, built by Port-Liner, 52 meters long and just under seven meters wide, will be able to carry twenty-four 20-foot containers weighing a total of 425 tons.[68] With that said, there is still no electronic equivalent to a large-scale shipping vessel capable of hauling 10,000 tons. Why? Once again, the batteries needed would be much too heavy.

In the mid-2000s, there was a lot of talk about hydrogen energy storage and hydrogen fuel cell cars. More recently this talk has died down. But perhaps we shouldn't be too hasty in dismissing hydrogen's role in the energy revolution. Why? Because we will need ships and planes that don't use fossil fuels. This, by extension, means electrifying them. Utilizing battery energy storage just isn't suitable for this purpose due to their weight. Hydrogen, on the other hand, is incredibly energy dense, packing three times more energy per kilogram than jet fuel. When hydrogen is burned inside of a fuel cell to produce electricity, its only byproduct is water so clean that you could drink it.[69]

*Hydrogen i*s the most plentiful element in the universe. Unfortunately for us, it doesn't occur naturally as a gas on earth. It is always combined with another element such as oxygen in the case of water and carbon in the case of fossil fuels. The vast majority of hydrogen is produced by steam reforming, a process in which high-temperature steam is used to produce hydrogen from natural gas.[70] Hydrogen is of paramount importance in oil refining as it acts as a necessary part of hydrodesulfurization, a process which removes sulfur from refined products made by the oil and gas industry, and hydrocracking, a chemical process which converts "heavier," less valuable, petroleum products into "lighter," more valuable products such as gasoline, jet fuel, and diesel.[71] While there is some ongoing research on using hydrogen produced from steam reforming in fuel cells, hydrogen can only be considered a sustainable method of energy storage if it is extracted in a manner that does not lead to more carbon emissions.

A more sustainable method to produce hydrogen is water electrolysis, which is when electricity is used to split water into hydrogen and oxygen. The sustainability of such a methodology is dependent on the source of energy used.

Within recent years a more novel approach for hydrogen production, called Microbial Electrolysis Cells (MECs), has captured the attention of researchers. MECs, to simplify it down substantially, "feed" electricity to bacteria, who in turn produce hydrogen. Up until recently, MECs only existed on laboratory benches. However, between 2013 and 2014, Elizabeth Heidrich of Newcastle University, whom I've had the pleasure of working under during my master's dissertation project in 2015, conducted a 12-month-long pilot-scale test on an MEC.[72] While only producing enough hydrogen to recoup half of the energy injected into it, a more robust design would bring up its efficiency substantially (trust me, I've seen the thing up close and worked for months to improve its efficiency, this technology has *potential.*).

While the aviation industry has yet to develop a viable commercial hydrogen fuel cell aircraft, a lot of work is currently being undertaken by the industry to make it happen. During the third quarter of 2016, German scientists successfully tested the world's first four-passenger hydrogen-fueled plane.[73] That same year, easyJet announced an initiative to utilize large-scale hybrid hydrogen fuel cell planes within a decade.[74] Physical trials are expected to begin in 2019.[75] Rolls-Royce, Airbus, and Siemens have already started the process of developing a commercial scale hybrid hydrogen fuel cell plane which they hope to test in 2020.[76] NASA is also currently in the process of developing its own hydrogen plane.[77]

While some might be concerned that the aviation industry, and for that matter the international shipping industry, hasn't yet developed non-carbon intensive transport systems that can be deployed right now, I don't believe that this is something we have to worry all that much about. This is because, when looked at in a global context, the emissions of the aviation and shipping industries pale in comparison to those of the automotive industry. Electrifying cars is, at the moment, far more critical than electrifying ships and planes. Moreover, the various technologies needed to produce hydrogen-fueled electric airplanes and shipping vessels already exist. We don't have to wait for them to be invented. The engineering problem is, how do we scale up these technologies for commercial use? While the aviation and shipping industries should not rest on their laurels (they already aren't), I believe that greater focus on these industries can be put off until the latter half of the century. In brief, batteries for cars, hydrogen for planes and ships. Electrify cars now. Electrify the rest later.

## Carbon From The Air

Most believe that transitioning our energy sources away from fossil fuels to renewables means the inevitable death of the fossil fuel industry. This is true up to a point. Coal is very much on its way to the grave, but, and many would find this to be a controversial position to take, I don't believe we can say the same for oil and gas. This, however, comes with one major caveat. We have to stop looking at them as fuel. Oil and gas are incredibly versatile substances, and in my opinion, the worst thing you can do with them is to burn them. Chemicals derived from crude oil alone can be found in aspirin, crayons, body lotion, rooftop waterproofing paint, the asphalt we build our roads from, the plastic coating on electrical wires, classical guitar strings, tires, and the list goes on ad nauseam.[78]

However, the amount of oil and gas that would need to be extracted for chemical use is much smaller than the amounts required for energy production. With that said, chemicals are in themselves far more profitable than the raw fossil fuels. For example, a barrel of crude oil currently costs about $70,[79] while a ton of paraxylene, a polyaromatic hydrocarbon used in polymers, is priced at over $1,000 per ton.[80] With the demand for petrochemicals increasing, it could very well be the case that the oil and gas industry of tomorrow will be just as profitable as it is today.

Processes related to the extraction, refining, and transport of fossil fuels account for about 10 percent of global emissions. This means that if the oil and gas industry is to make the transition from the energy business to the chemical business, it must find a way to deal with its carbon emissions. With carbon capture and storage (CCS), the carbon dioxide produced in a combustion process is drawn off (captured) before it can be emitted into the atmosphere. The captured carbon dioxide is then placed under high pressure and injected into a suitable rock formation (stored). If the process of using fossil fuels for energy is the extraction of carbon from the earth's crust and dispersing it into the atmosphere, CCS is its opposite, extracting carbon from fossil fuels and putting them back into the earth.

The idea of capturing carbon dioxide from fossil fuel combustion processes forces us to ask a question of pivotal importance. Is it not feasible for us to produce technology that removes carbon dioxide from the air? On a purely technological level, we already possess the means to pull carbon from the atmosphere. In the third quarter of 2018, a Canadian company called Carbon Engineering published a breakthrough paper demonstrating that carbon could be pulled out of the atmosphere at a cost of $100 per ton. At this rate, it would currently cost over $3 trillion to remove one year's worth of emissions from the atmosphere.[81]

The idea of so-called "negative emissions" technologies has enamored many scientists for the better part of two decades. The belief is that not only could such technologies slow down global warming, but they might also reverse it. Before Carbon Engineering's paper, the idea of pulling carbon from the air, while in existence, hadn't been compelling due to the exorbitant cost. Previous studies estimated the cost of removing a ton of carbon to be around $600, meaning that removing a year's worth of emissions would cost just under $20 trillion, equivalent to two-thirds of the world's gross domestic product.

The problem is, however, that while everything looks good on paper, no one has as of yet convincingly *physically* demonstrated that carbon can be pulled from the atmosphere at the scale necessary to make a dent. This is worrying as the technology is massively overhyped. Of the 116 models used by the IPCC to limit warming to 2°C, 101 relied on the implementation of negative emission technologies at a global scale.[82] With the exception of hydrogen-fueled commercial airliners and shipping vessels, every technology analyzed in this chapter has so far been put into practice and proven its reliability. Placing such focus and hope on a technology that has not demonstrated itself is not a good idea. Assuming that it might be one day feasible for us to undo the damage we've done by pulling carbon out of the air remains a high stakes gamble.

This does not mean that negative emissions technologies shouldn't be pursued. They should. And with great vigor. However, it is not wise to put our total faith in such technology. If atmospheric carbon capture develops into a technology that we can use to reverse warming (as I do believe it eventually will), then great. But if it doesn't, the technologies needed to mitigate our emissions right now already exist. It is in them that we should place our faith.

## Impact

To know that the technology needed to mitigate climate change, for the most part, already exists is both comforting and infuriating. It is comforting because it allows us to know that this is a problem that we can solve. It is infuriating as not enough action is being taken to deploy the technologies mentioned earlier at the sort of scale necessary.

It is important here to address the elephant in the room. Limiting global warming to 2°C does not mean that the problem is solved. Mitigation must ultimately be coupled with adaptation. That being said, adaptation will be painfully tedious, complex, challenging, and virtually impossible to do preemptively due to the uncertainties at play. Relegating ourselves to adaptation means reacting rather than acting. Every dollar spent on mitigation serves as many dollars saved in adaptation costs.

Reengineering our agricultural sector to be more sustainable is not a very costly affair, and it may even prove to save farmers and agriculturalists money. On the other hand, transitioning our energy infrastructure to be completely renewable by 2050 will require a significant financial investment. The International Renewable Energy Agency estimates that investments in our energy infrastructure will need to increase by 30 percent between 2015 and 2050, from $93 trillion (what we already plan to spend) to $120 trillion. "In total, throughout the period, the global economy would need to invest around 2% of the average global GDP per year in decarbonization solutions, including renewable energy, energy efficiency, and other enabling technologies."[83]

It is essential to appreciate that money spent on mitigation is an investment. It is an investment in a more robust and reliable energy system, in cleaner air and water, in better and more affordable transport, in a more productive agricultural sector. It is estimated that we will need to spend an additional 1.7 trillion dollars, globally, every year, to transition to a renewable future. This investment is expected to save the world economy at least 6 trillion dollars annually due to fewer healthcare and environmental costs. On the other hand, money spent on adaptation is purely cost. It is effort and money put into maintaining our current infrastructure. There is no payback.

Given the enormity of the challenge we face, we must ask ourselves, how will history remember us if we should fail? What will be our impact? Will we limit warming to 2°C, something that is well within our grasp, and adapt as far as we can, or will we allow billions to undergo unnecessary suffering? To quote John Holdren, "We basically have three choices: mitigation, adaptation and suffering. We're going to do some of each. The question is what the mix is going to be. The more mitigation we do, the less adaptation will be required and the less suffering there will be."[84]

# CONCLUSION

*I am an optimist. It does not seem too much use being anything else.*

    *– Winston Churchill*

With humans enjoying more financially prosperous, healthier, and more peaceful lives than at any prior point in human history, with thousands being pulled out of abject poverty every day, the question we face is; is this upward trajectory sustainable?

We, humans, are a resilient and resourceful bunch, and we've become exceedingly good at solving the problems we create. It is not fatalism nor self-loathing that will get us out of this mess. Rather, it is pragmatism and the ability to see the issue as it is. Climate change is quite possibly the greatest problem humanity has ever faced. If it is not dealt with properly, it has the potential to bring with it a great deal of unnecessary suffering upon billions of people. Despite its scale, climate change is a problem that we can tackle head on through the use of technology and suitable policies.

Many assume that our carbon emissions can be mitigated through behavioral changes. While there are indeed many ways that a person can reduce their individual carbon footprint, such as using less electricity, opting to walk or take the bus rather than use a car, eating a plant-based diet, and living a zero-waste lifestyle, the fact of the matter is that such actions, while noble and commendable, don't create that much of an impact. This is because, at its heart, the problem of climate change is a problem of technology, or should I say specific technologies. Within our carbon-intensive paradigm, behavioral changes may have a slight benefit. However, the problem simply evaporates the instant you replace a carbon-intensive technology, say burning coal to produce electricity, with the carbon neutral alternative, say solar panels and wind turbines.

Many have argued that mitigating climate change is something that isn't economically feasible. Not true. Reforming our agricultural practices to emit less carbon should cost the world economy close to nothing, and might actually save farmers money. Meanwhile, moving our entire energy infrastructure to renewables such as wind, solar and hydro by 2050 only requires investing 2 percent of global GDP annually (we already annually spend 6 percent of global GDP on indirect fossil fuel subsidies).

Deploying such technologies at a global scale, within the necessary timeframe, can only be achieved with suitable economic and political policies. Unfortunately, the international community has failed time and time again to put in the necessary safeguards that would allow such a transformation to occur.

The Paris Climate Accords provided us with a ray of light in late 2015. However, critical flaws in the treaty, such as not taking into account the idea of a global carbon budget and having no enforcement mechanism, means that even if every signatory of the Accord met its 2015 pledges, which is unlikely, warming would still exceed 3°C by the end of the century. Policymakers have failed time and time again to bring into being the sorts of policies necessary to mitigate climate change. As such, it has become the responsibility of the general public to pressurize policymakers into action.

From the civil rights movement in America during the 1960s, to Gandhi protesting the British rule of India, to the Suffragettes who fought to earn women the right to vote in many countries in the late 19th and early 20th centuries, grassroots activism is the path that many throughout history have taken when all other doors are shut. The nexus of politics, economics, and the environment have one thing in common: the voice of the people.

The modern environmental movement was born on the 22nd of April 1970. Riding on the coattails of the civil rights and anti-war movements of the 1960s, Earth Day was the world's first environmental mass movement driven by the grassroots, with more than 20 million people in attendance. At the time there was no such thing as an environmental regulator, no laws or guidelines to protect air and water quality. The momentum gained by the environmental movement built the necessary pressure which drove Richard Nixon, hardly a tree hugger, to found the United States Environmental Protection Agency (EPA). The EPA would soon create the Clean Air Act and the Clean Water Act, as well as outlaw the use of DDT and asbestos. The environmental movement, which found its origins in American counterculture, would soon make its way all over the globe.

The awareness raised by the movement, which in reality was made up of many smaller movements working in tandem towards a higher goal, created the kind of pressure necessary to usher in a litany of environmental laws, guidelines, regulations and international treaties. In 1987 the international community came together to establish the Montreal Protocol, which aimed to phase out chemicals that have an adverse impact on the world's ozone layer. Two years later, they would come together once more to sign the Basel Convention, which prohibited the transport of hazardous waste across borders.

Today, a myriad of grassroots environmental and climate-related advocacy groups work in tandem, both on the local and on the international scale, to bring about the sort of change needed to face this problem. 350.org, which gets its name from the belief that atmospheric concentrations of carbon should be limited to 350ppm (they're currently over 400ppm), utilizes "online campaigns, grassroots organizing, and mass public actions to oppose new coal, oil and gas projects, take money out of the companies that are heating up the planet, and build 100% clean energy solutions that work for all."[1] The Sierra Club's Beyond Coal campaign aims to "replace dirty coal with clean energy by mobilizing grassroots activists in local communities to advocate for the retirement of old and outdated coal plants and to prevent new coal plants from being built."[2] The Climate Action Network is a "worldwide network of over 1300 Non-Governmental Organizations (NGOs) in more than 120 countries, working to promote government and individual action to limit human-induced climate change to ecologically sustainable levels."[3]

Despite these efforts, within recent years, grassroots climate change activism has failed to gather the sort of momentum needed to place adequate pressure on policymakers. Some think that this is because climate advocacy groups have become far too politicized, mired in infighting, and focused on the lowest common denominator, i.e., proving climate change denialists wrong. While there's most certainly some truth in this analysis, I don't believe this paints a full picture.

Environmental activism has accomplished a great deal within the last half-century. That being said, climate change is no mere environmental problem. It is a wicked environmental, economic, political, and social problem that can only be tackled through the rapid deployment of carbon-neutral alternatives at a breakneck pace. The burden of creating the sort of pressure required to force the hand of policymakers cannot remain only upon the heads of environmental activists. Anyone who understands the severity of the issue has a moral responsibility to draw attention towards it. This burden now falls to you.

I began this book by stating that despite the numerous challenges facing humanity today, we are living in what is by most accounts the most prosperous period in human history. If we don't take climate change seriously, our children won't be able to say the same.

# NOTES & RESOURCES

## Introduction

1. "Global Extreme Poverty." Our World In Data, https://ourworldindata.org/extreme-poverty.
2. "War And Peace." Our World In Data, 2018, https://ourworldindata.org/war-and-peace.
3. "Life Expectancy." Our World In Data, 2018, https://ourworldindata.org/life-expectancy.

## Chapter 1: The Theory

1. Harari, Yuval N. Sapiens. 2014.
2. Fleming, James Rodger. Historical Perspectives On Climate Change. Oxford University Press, 2005.
3. Weart, Spencer R. The Discovery Of Global Warming. Harvard University Press, 2009.
4. Archer, David. Global Warming: Understanding The Forecast. 2006.

5. "Svante Arrhenius : Feature Articles". Earthobservatory.Nasa.Gov, https://earthobservatory.nasa.gov/Features/Arrhenius/arrhenius_3.php.

6. McKibben, Bill. Deep Economy. Henry Holt & Company, 2008.

7. Serway, Raymond A. Physics For Scientists & Engineers, With Modern Physics. Saunders College Pub., 1990.

8. Romm, Joseph J. Climate Change: What Everyone Needs To Know. 2015.

9. "The Sun's Evolution". Faculty.Wcas.Northwestern.Edu, 2018, http://faculty.wcas.northwestern.edu/~infocom/The%20Website/evolution.html.

10. "Milankovitch Cycles And Glaciation". Indiana.Edu, http://www.indiana.edu/~geol105/images/gaia_chapter_4/milankovitch.htm.

11. Goosse H., P.Y. Barriat, W. Lefebvre, M.F. Loutre and V. Zunz, (2008-2010). Introduction to climate dynamics and climate modeling. Online textbook available at http://www.climate.be/textbook.

12. "Data.GISS: GISS Surface Temperature Analysis: Analysis Graphs And Plots". Data.Giss.Nasa.Gov, https://data.giss.nasa.gov/gistemp/graphs_v3/.

13. "Long-Term Global Warming Trend Continues : Image Of The Day". Earthobservatory.Nasa.Gov, https://earthobservatory.nasa.gov/IOTD/view.php?id=80167. Accessed 24 May 2018.

14. "Solar Radiation And Climate Experiment (SORCE) Fact Sheet : Feature Articles". Earthobservatory.Nasa.Gov, https://earthobservatory.nasa.gov/Features/SORCE/.

15. Cook, John. "The CO2/Temperature Correlation Over The 20Th Century". Skeptical Science, 2009, https://www.skepticalscience.com/The-CO2-Temperature-correlation-over-the-20th-Century.html.

16. Team, ESRL. "ESRL Global Monitoring Division - Education And Outreach". Esrl.Noaa.Gov, https://www.esrl.noaa.gov/gmd/outreach/isotopes/.

17. "What Is Causing The Increase In Atmospheric CO2?". Skeptical Science, https://skepticalscience.com/print.php?r=384.

**Chapter 2: Gazing Into The Crystal Ball**

1. Burry, Michael. "Opinion I I Saw The Crisis Coming. Why Didn't The Fed?". Nytimes.Com, 2010, https://www.nytimes.com/2010/04/04/opinion/04burry.html. Accessed 1 Oct 2018.

2. Harris, Mark et al. "Inside The First Church Of Artificial Intelligence I Backchannel". WIRED, https://www.wired.com/story/anthony-levandowski-artificial-intelligence-religion/.

3. "Top 10 Bad Tech Predictions". Digital Trends, https://www.digitaltrends.com/features/top-10-bad-tech-predictions/5/.

4. Allen, Frederick E. "The Myth Of Artificial Intelligence I AMERICAN HERITAGE". Americanheritage.Com, 2001, https://www.americanheritage.com/content/myth-artificial-intelligence.

5. "Convert Acceleration: 9.80665 M/S2 (Meter Per Second Squared) To ...". Convert-Units.Info,

http://convert-units.info/acceleration/meter-second2/9.80665.

6. "Gravity Of Earth". En.M.Wikipedia.Org, 2018, https://en.m.wikipedia.org/wiki/Gravity_of_Earth.

7. "Instant Expert: General Relativity". New Scientist, 2018, https://www.newscientist.com/round-up/instant-expert-general-relativity/.

8. Buchen, Lizzie. "May 29, 1919: A Major Eclipse, Relatively Speaking". WIRED, 2009, https://www.wired.com/2009/05/dayintech-0529/.

9. "Hafele–Keating Experiment". En.M.Wikipedia.Org, 2018, https://en.m.wikipedia.org/wiki/Hafele–Keating_experiment.

10. Root, Terry L. "Why We Need Climate Models". Audubon, 2014, https://www.audubon.org/magazine/september-october-2014/why-we-need-climate-models.

11. "IPCC - Intergovernmental Panel On Climate Change". Ipcc.Ch, 2001, http://www.ipcc.ch/ipccreports/tar/vol4/index.php?idp=91. Accessed 1 Oct 2018.

12. Scherer, Glenn. "How The IPCC Underestimated Climate Change". Scientific American, 2012, https://www.scientificamerican.com/article/how-the-ipcc-underestimated-climate-change/.

13. Carrington, Damian. "IPCC Officials Admit Mistake Over Melting Himalayan Glaciers". The Guardian, 2010, https://www.theguardian.com/environment/2010/jan/20/ipcc-himalayan-glaciers-mistake.

14. Burch, John R. Climate Change And American Policy: Key Documents, 1979–2015. 2016.

15. "Climate Communication | Heat Waves: The Details". Climatecommunication.Org, https://www.climatecommunication.org/new/features/heat-waves-and-climate-change/heat-waves-the-details/.

16. Majeed, Aamir, and Aamir Majeed. "Mortuaries Fill Up As Heat Continues To Take Scalps". Pakistantoday.-Com.Pk, 2015, https://www.pakistantoday.com.pk/2015/06/22/mortua ries-fill-up-as-heat-continues-to-take-scalps/.

17. Jorgic, Drazen, and Syed Raza Hassan. "Pakistan City Readies Graves, Hospitals, In Case Heat Wave Hits Again". Reuters, 2016, https://www.reuters.com/article/us-pakistan-heatwave-idUSKCN0YB0TU.

18. Astor, Maggie. "Hottest April Day Ever Was Probably Monday In Pakistan: A Record 122.4°F". Nytimes.-Com, 2018, https://www.nytimes.com/2018/05/04/world/asia/pak istan-heat-record.html.

19. "Impacts Of Summer 2003 Heat Wave In Europe". https://www.unisdr.org/files/1145_ewheatwave.en.pdf.

20. "Record-Breaking 2010 Eastern European/Russian Heatwave". Sciencedaily, 2011, https://www.sciencedaily.com/releases/2011/03/110318 091141.htm. Accessed 2 Oct 2018.

21. "Record-Breaking 2010 Heatwave". Ethlife.Ethz.Ch, 2011, http://www.ethlife.ethz.ch/archive_articles/110318_Hitz ewelle_2010_MM/index_EN.html.

22. Herring, S. C., N. Christidis, A. Hoell, J. P. Kossin, C. J. Schreck III, and P. A. Stott, Eds., 2018: Explaining Extreme Events of 2016 from a Climate Perspective. Bull. Amer. Meteor. Soc., 99 (1), S1–S157.

23. Zurayk, Rami. "Use Your Loaf: Why Food Prices Were Crucial In The Arab Spring". The Guardian, 2011, https://www.theguardian.com/lifeandstyle/2011/jul/17/bread-food-arab-spring.

24. Eplett, Layla. "What's An Aleppo Pepper?". Scientific American Blog Network, 2016, https://blogs.scientificamerican.com/food-matters/what-s-an-aleppo-pepper/.

25. Gleick, Peter H. "Water, Drought, Climate Change, And Conflict In Syria". American Meteorological Society, 2014, https://journals.ametsoc.org/doi/abs/10.1175/WCAS-D-13-00059.1.

26. U.S. House Of Representatives Document Repository, 2018, https://docs.house.gov/billsthisweek/20171113/HRPT-115-HR2810.pdf. Accessed 4 Aug 2018.

27. Trenberth, KE. "Changes In Precipitation With Climate Change". Climate Research, vol 47, no. 1, 2011, pp. 123-138. Inter-Research Science Center, doi:10.3354/cr00953.

28. United States Department Of Agriculture Economic Research Service, 2012, https://www.ers.usda.gov/topics/in-the-news/us-drought-2012-farm-and-food-impacts.aspx#.VDMwNvldVrU. Accessed 3 June 2018.

29. Mooney, Chris. "Wildfires Used To Be Rare In The

Great Plains. They'Ve More Than Tripled In 30 Years". Washington Post, 2017, https://www.washingtonpost.com/news/energy-environment/wp/2017/06/16/scientists-find-a-400-percent-increase-in-wildfire-destruction-in-the-great-plains/.

30. "Extreme Precipitation And Climate Change". Center For Climate And Energy Solutions, https://www.c2es.org/content/extreme-precipitation-and-climate-change/.

31. "Explainer: What Climate Models Tell Us About Future Rainfall | Carbon Brief". Carbon Brief, https://www.carbonbrief.org/explainer-what-climate-models-tell-us-about-future-rainfall.

32. "World Population Growth". Our World In Data, https://ourworldindata.org/world-population-growth.

33. "World Population Clock: 7.7 Billion People (2018) - Worldometers". Worldometers.Info, 2018, http://www.worldometers.info/world-population/. Accessed 19 Sept 2018.

34. "World Population Projected To Reach 9.8 Billion In 2050, And 11.2 Billion In 2100 | UN DESA | United Nations Department Of Economic And Social Affairs". UN DESA | United Nations Department Of Economic And Social Affairs, 2017, https://www.un.org/development/desa/en/news/population/world-population-prospects-2017.html.

35. American Geophysical Union. "Groundwater Resources Around The World Could Be Depleted By 2050S". Phys.Org, 2016, https://phys.org/news/2016-12-groundwater-resources-world-depleted-2050s.html.

36. "CLIMATE CHANGE AND HEALTH". World Bank, https://www.worldbank.org/en/topic/climatechangean dhealth.

37. "Competing For Clean Water Has Led To A Crisis". Nationalgeographic.Com, https://www.nationalgeographic.com/environment/fres hwater/freshwater-crisis/.

38. "Climate Impacts On Water Resources I Climate Change Impacts I US EPA". 19January2017snapshot.Epa.-Gov, 2017, https://19january2017snapshot.epa.gov/climate-impacts/climate-impacts-water-resources_.html.

39. "Carbon Dioxide And Carbonic Acid". Ion.Chem.Usu.Edu, http://ion.chem.usu.edu/~sbialkow/Classes/3650/Car-bonate/Carbonic%20Acid.html. Accessed 2 Oct 2018.

40. Romm , Joseph J. Climate Change: What Everyone Needs To Know. 2015.

41. "Percent Change In Acidity". Pmel.Noaa.Gov, https://www.pmel.noaa.gov/co2/file/Percent+change+i n+acidity.

42. "What Is Ocean Acidification?". Pmel.Noaa.Gov, https://www.pmel.noaa.gov/co2/story/What+is+Ocean +Acidification%3F.

43. "Earth Matters - Mollusks, Corals, Carbon, And Volcanoes". Earthobservatory.Nasa.Gov, 2012, https://earthobservatory.nasa.gov/blogs/earthmatters/2 012/05/30/mollusks-corals-carbon-and-volcanoes/.

44. "Coral Reefs". WWF Global,

http://wwf.panda.org/our_work/oceans/coasts/coral_r eefs/.

45. Crabbe, M. James C. "Climate Change, Global Warming And Coral Reefs: Modelling The Effects Of Temperature". Computational Biology And Chemistry, vol 32, no. 5, 2008, pp. 311-314. Elsevier BV, doi:10.1016/j.-compbiolchem.2008.04.001.

46. Ortiz, Erik. ""Trouble In Paradise": Why Death Of World's Coral Is Alarming". NBC News, 2018, https://www.nbcnews.com/news/world/scope-great-barrier-reef-s-massive-coral-bleaching-alarms-scientists-n867521.

47. Harvey, Chelsea. "Recent Ocean Heat Waves Have "Forever" Altered Great Barrier Reef". Scientific American, 2018, https://www.scientificamerican.com/article/recent-ocean-heat-waves-have-forever-altered-great-barrier-reef/.

48. Becatoros, Elena. "More Than 90 Percent Of Coral Reefs Will Die Out By 2050". The Independent, 2017, https://www.independent.co.uk/environment/environ ment-90-percent-coral-reefs-die-2050-climate-change-bleaching-pollution-a7626911.html.

49. Carilli, Jessica. "Why Are Coral Reefs Important?". Nature.Com, 2013, https://www.nature.com/scitable/blog/saltwater-science/why_are_coral_reefs_important.

50. Doll, Jen. "How Storms Got Their Names". The Atlantic, 2012, https://www.theatlantic.com/international/archive/201 2/08/how-storms-got-their-names/324163/.

51. "Tropical Cyclone Naming History And Retired Names". Nhc.Noaa.Gov, https://www.nhc.noaa.gov/aboutnames_history.shtml.

52. "NHC Estimates Harvey Damage At $125 Billion -- Occupational Health & Safety". Occupational Health & Safety, 2018, https://ohsonline.com/articles/2018/01/29/nhc-estimates-harvey-damage.aspx. Accessed 2 Oct 2018.

53. Chappell, Bill. "NPR Choice Page". Npr.Org, 2017, https://www.npr.org/sections/thetwo-way/2017/08/28/546776542/national-weather-service-adds-new-colors-so-it-can-map-harveys-rains. Accessed 2 Oct 2018.

54. AMADEO, KIMBERLY. "Hurricane Harvey Shows How Climate Change Can Impact The Economy". The Balance, 2018, https://www.thebalance.com/hurricane-harvey-facts-damage-costs-4150087.

55. Miller, Brandon. "All The Records Irma Has Already Broken". CNN, 2017, https://edition.cnn.com/2017/09/10/us/irma-facts-record-numbers-trnd/index.html.

56. "5.6 Million Floridians Ordered To Evacuate As Irma Closes In". Www.Newvision.Co.Ug, 2017, https://www.newvision.co.ug/new_vision/news/1461263/56-million-floridians-evacuate-irma-closes.

57. Ingraham, Christopher. Washington Post, 2018, https://www.washingtonpost.com/news/wonk/wp/2018/06/02/hurricane-maria-was-one-of-the-deadliest-natural-disasters-in-u-s-history-according-to-a-new-estimate/?noredirect=on&utm_term=.f4f1bf52096b. Accessed 2 Oct 2018.

58. Fink, Sheri. "Puerto Rico'S Hurricane Maria Death Toll Could Exceed 4,000, New Study Estimates". Nytimes.-Com, 2018, https://www.nytimes.com/2018/05/29/us/puerto-rico-deaths-hurricane.html.

59. "Hurricanes And Climate Change". Union Of Concerned Scientists, 2017, https://www.ucsusa.org/global-warming/science-and-impacts/impacts/hurricanes-and-climate-change.html#.W4Q3sZMzYdo. Accessed 7 Aug 2018.

60. "Global Warming And Hurricanes: An Overview Of Current Research Results". Geophysical Fluid Dynamics Laboratory, 2018, https://www.gfdl.noaa.gov/global-warming-and-hurricanes/. Accessed 2 Oct 2018.

61. Woodward, Aylin. "Climate Change Blamed For Arabian Sea'S Unexpected Hurricanes". New Scientist, 2017, https://www.newscientist.com/article/climate-change-blamed-arabian-seas-unexpected-hurricanes/.

62. Tropical Cyclone "Mekunu" slams into Oman as strongest in recorded history, dumps nearly 3 years' worth of rain in one day. (2018). Retrieved from https://watchers.news/2018/05/26/tropical-cyclone-mekunu-oman-yemen/

63. Cockburn, H. (2018). Coral reef growth 'already failing to keep pace with sea level rise', study says. Retrieved from https://www.independent.co.uk/environment/coral-reef-sea-level-rise-climate-change-global-warming-great-barrier-a8397591.html

64. Fletcher, S. (2015). Causes of Sea Level Rise.

Retrieved from http://docplayer.net/246177-Causes-of-sea-level-rise.html

65. What Are the Range of Possibilities for Sea Level Rise Projections?. (2014). Retrieved from https://www.ucsusa.org/publications/ask/2014/sea-level-rise.html#.W4REOZMzYdo

66. Hansen, J., Sato, M., Hearty, P., Ruedy, R., Kelley, M., & Masson-Delmotte, V. et al. (2016). Ice melt, sea level rise and superstorms: evidence from paleoclimate data, climate modeling, and modern observations that 2 °C global warming could be dangerous. Atmospheric Chemistry And Physics, 16(6), 3761-3812. doi: 10.5194/acp-16-3761-2016

67. Sea-Ice Basics. Retrieved from https://www.ldeo.columbia.edu/louisab/sedpage/basics

68. Watts, Jonathan, and Josh Holder. "Arctic'S Strongest Sea Ice Breaks Up For First Time On Record". The Guardian, 2018, https://www.theguardian.com/world/2018/aug/21/arctics-strongest-sea-ice-breaks-up-for-first-time-on-record.

69. "Quick Facts On Ice Sheets | National Snow And Ice Data Center". Nsidc.Org, https://nsidc.org/cryosphere/quickfacts/icesheets.html.

70. Watts, Jonathan, and Josh Holder. "Arctic'S Strongest Sea Ice Breaks Up For First Time On Record". The Guardian, 2018, https://www.theguardian.com/world/2018/aug/21/arctics-strongest-sea-ice-breaks-up-for-first-time-on-record.

71. "Greenland Ice Sheet Surface Mass Budget: DMI". Dmi.Dk,

https://www.dmi.dk/en/groenland/maalinger/greenla nd-ice-sheet-surface-mass-budget/.

72. "Polar Ice Sheets Melting Faster Than Ever". DW.COM, 2013, https://www.dw.com/en/polar-ice-sheets-melting-faster-than-ever/a-16432199.

73. "Mass Balance Of The Antarctic Ice Sheet From 1992 To 2017". Vol 558, no. 7709, 2018, pp. 219-222. Springer Nature, doi:10.1038/s41586-018-0179-y.

74. Zscheischler, Jakob et al. "Future Climate Risk From Compound Events". Nature Climate Change, vol 8, no. 6, 2018, pp. 469-477. Springer Nature, doi:10.1038/s41558-018-0156-3.

75. Abou-Hadid, Ayman F. AFED 2014 Report. 2014, p. Chapter 5.

76. Hereher, Mohamed E. "Vulnerability Of The Nile Delta To Sea Level Rise: An Assessment Using Remote Sensing". Geomatics, Natural Hazards And Risk, vol 1, no. 4, 2010, pp. 315-321. Informa UK Limited, doi:10.1080/19475705.2010.516912.

77. "In Egypt, A Rising Sea — And Growing Worries About Climate Change's Effects". Npr.Org, 2017, https://www.npr.org/sections/parallels/2017/08/13/54 2645647/in-egypt-a-rising-sea-and-growing-worries-about-climate-changes-effects.

78. Elsharkawy, H et al. "Climate Change: The Impacts Of Sea Level Rise On Egypt". 45Th ISOCARP Congress, 2009, https://orca.cf.ac.uk/44506/3/ISOCARP_2009_Impact%2 0of%20Sea%20Level%20Rise%20on%20Egypt.pdf.

79. Kishore, Nishant et al. "Mortality In Puerto Rico After Hurricane Maria". New England Journal Of Medi-

cine, vol 379, no. 2, 2018, pp. 162-170. New England Journal Of Medicine (NEJM/MMS), doi:10.1056/nejm-sa1803972.

80. Brosha, Stephen. "The Environment And Conflict In The Rwandan Genocide". Atlismta.Org, https://atlismta.org/online-journals/0607-journal-development-challenges/the-environment-and-conflict-in-the-rwandan-genocide/.

81. "NATIONAL SECURITY IMPLICATIONS OF CLIMATE-RELATED RISKS AND A CHANGING CLIMATE". 2015, https://archive.defense.gov/pubs/150724-congressional-report-on-national-implications-of-climate-change.pdf?source=govdelivery.

82. "Hyperthermals: What Can They Tell Us About Modern Global Warming? | Carbon Brief". Carbon Brief, 2017, https://www.carbonbrief.org/hyperthermals-what-can-they-tell-us-about-modern-global-warming.

83. Anderson, Bob. "Does Temperature Control Atmospheric Carbon Dioxide Concentrations?". State Of The Planet, 2017, https://blogs.ei.columbia.edu/2010/07/07/does-temperature-control-atmospheric-carbon-dioxide-concentrations/. Accessed 8 Oct 2018.

84. MCSWEENEY, R. (2015). Amazon rainforest is taking up a third less carbon than a decade ago | Carbon Brief. [online] Carbon Brief. Available at: https://www.carbonbrief.org/amazon-rainforest-is-taking-up-a-third-less-carbon-than-a-decade-ago [Accessed 9 Jul. 2018].

85. "Methane And Frozen Ground | National Snow

And Ice Data Center". Nsidc.Org, https://nsidc.org/cryosphere/frozenground/methane.ht ml.

86. Ruppel, Carolyn D., and John D. Kessler. "The Interaction Of Climate Change And Methane Hydrates". Reviews Of Geophysics, vol 55, no. 1, 2017, pp. 126-168. American Geophysical Union (AGU), doi:10.1002/2016rg000534.

87. Robock, Alan. "Could Subsea Methane Hydrates Be A Warming "Tipping Point"? - Eos". Eos, 2017, https://eos.org/editors-vox/could-subsea-methane-hydrates-be-a-warming-tipping-point.

88. Öffentlichkeitsarbeit, Medien-. "Thawing Permafrost Produces More Methane Than Expected". Uni-Hamburg.De, 2017, https://www.uni-hamburg.de/en/newsroom/presse/2018/pm18.html.

## Chapter 3: Conspiracy, Deceit & Defamation of Character

1. Banerjee, Neela et al. "Exxon's Own Research Confirmed Fossil Fuels' Role In Global Warming Decades Ago". Insideclimate News, 2015, https://insideclimatenews.org/news/15092015/Exxons-own-research-confirmed-fossil-fuels-role-in-global-warming.

2. Banerjee, Neela et al. "Exxon Believed Deep Dive Into Climate Research Would Protect Its Business". Inside-climate News, 2015, https://insideclimatenews.org/news/16092015/exxon-believed-deep-dive-into-climate-research-would-protect-

its-business.

3. Song, Lisa et al. "Exxon Confirmed Global Warming Consensus In 1982 With In-House Climate Models". Insideclimate News, 2015, https://insideclimatenews.org/news/18092015/exxon-confirmed-global-warming-consensus-in-1982-with-in-house-climate-models.

4. Times, Philip. "Global Warming Has Begun, Expert Tells Senate". Nytimes.Com, 1988, https://www.nytimes.com/1988/06/24/us/global-warming-has-begun-expert-tells-senate.html.

5. Banerjee, Neela. "Exxon's Oil Industry Peers Knew About Climate Dangers In The 1970S, Too". Insideclimate News, 2015, https://insideclimatenews.org/news/22122015/exxon-mobil-oil-industry-peers-knew-about-climate-change-dangers-1970s-american-petroleum-institute-api-shell-chevron-texaco.

6. Shulman, Seth. "Smoke, Mirrors & Hot Air: How Exxonmobil Uses Big Tobacco'S Tactics To Manufacture Uncertainty On Climate Science". Union Of Concerned Scientists, 2007, https://www.ucsusa.org/sites/default/files/legacy/assets/documents/global_warming/exxon_report.pdf.

7. "Daily Doc: The "Frank Statement" Of 1954". Tobacco.Org, 2000, http://www.tobacco.org/Documents/dd/ddfrankstatement.html.

8. Readfearn, Graham. "Doubt Over Climate Science Is A Product With An Industry Behind It | Graham Readfearn". The Guardian, 2015,

https://www.theguardian.com/environment/planet-oz/2015/mar/05/doubt-over-climate-science-is-a-product-with-an-industry-behind-it.

9. Revkin, Andrew. "Industry Ignored Its Scientists On Climate". Nytimes.Com, 2009, https://www.nytimes.com/2009/04/24/science/earth/24deny.html?mtrref=undefined&gwh=3C507C3AD0B3FA645DE0929522BCF1CD&gwt=pay.

10. Howlett, Peter. How Well Do Facts Travel?. Cambridge University Press, 2011.

11. Goldstein, Natalie, and Kerry Harrison Cook. Global Warming. Facts On File, 2009.

12. "Climate Science Vs. Fossil Fuel Fiction". 2015, https://www.ucsusa.org/sites/default/files/attach/2015/03/APIquote1998_1.pdf.

13. Bardhan, Nilanjana, and C. Kay Weaver. Public Relations In Global Cultural Contexts. Routledge, 2011.

14. RAMSAY, ADAM. "Greenpeace Investigation Exposes Climate Denier Academics For Sale". Opendemocracy, 2015, https://www.opendemocracy.net/uk/adam-ramsay/greenpeace-investigation-exposes-climate-denier-academics-for-sale.

15. Carter, Lawrence, and Maeve McClenaghan. "Exposed: Academics-For-Hire Agree Not To Disclose Fossil Fuel Funding". Unearthed/Greenpeace, 2015, https://unearthed.greenpeace.org/2015/12/08/exposed-academics-for-hire/.

16. "What Do The 'Climategate' Hacked CRU Emails Tell Us?". Skeptical Science,

https://skepticalscience.com/Climategate-CRU-emails-hacked.htm.

17. "Awareness, Opinions About Global Warming Vary Worldwide". Gallup.Com, 2009, https://news.gallup.com/poll/117772/Awareness-Opinions-Global-Warming-Vary-Worldwide.aspx.

18. Steiner, Rick. "As Governor, Sarah Palin Believed In Climate Change". News Miner, 2016, http://www.newsminer.com/opinion/community_persp ectives/as-governor-sarah-palin-believed-in-climate-change/article_a5a7326c-1e36-11e6-8f8c-e7323ac76b06.html.

19. Steiner, Rick. "Before She Denied Climate Change, Sarah Palin Acknowledged And Confronted It". Anchorage Daily News, 2016, https://www.adn.com/commentary/article/she-denied-climate-change-sarah-palin-acknowledged-and-confronted-it/2016/05/16/.

20. Davenport, Coral, and Eric Lipton. "How G.O.P. Leaders Came To View Climate Change As Fake Science". Nytimes.Com, 2017, https://www.nytimes.com/2017/06/03/us/politics/rep ublican-leaders-climate-change.html.

21. Senecah, Susan. The Environmental Communication Yearbook: Volume 1. Routledge, 2013.

22. "The White House And The Greenhouse". Nytimes.Com, 1989, https://www.nytimes.com/1989/05/09/opinion/the-white-house-and-the-greenhouse.html.

23. Mims, Christopher. "Mitt Romney, Political Windsock, Flips To Climate Change Denial". Grist, 2011,

https://grist.org/article/2011-10-28-mitt-romney-political-windsock-flips-to-climate-change-denial/.

24. KAPLAN, REBECCA, and ELLEN UCHIMIYA. "Where The 2016 Republican Candidates Stand On Climate Change". CBS, 2016, https://www.cbsnews.com/news/where-the-2016-republican-candidates-stand-on-climate-change/.

25. "Fossil Fuel Funding Of 2016 Presidential Candidates". Greenpeace USA, 2016, https://www.greenpeace.org/usa/campaign-updates/fossil-fuel-funding-presidential-candidates/.

26. Boren, Zach. "US Election: Who Are The Most Fossil Fuel Funded Presidential Candidates?". Unearthed, 2016, https://unearthed.greenpeace.org/2016/03/03/us-election-who-are-the-most-fossil-fuel-funded-presidential-candidates/.

27. Wilson, Rachel. "Fossil Fuel Giants Poured Millions Into Trump'S Inauguration. Now It'S Paying Off.". Grist, 2017, https://grist.org/article/fossil-fuel-giants-poured-millions-into-trumps-inauguration-now-its-paying-off/.

28. Zhang, Hai-Bin et al. "U.S. Withdrawal From The Paris Agreement: Reasons, Impacts, And China's Response". Advances In Climate Change Research, vol 8, no. 4, 2017, pp. 220-225. Elsevier BV, doi:10.1016/j.accre.2017.09.002.

29. Nuccitelli, Dana. "Trump Just Cemented His Legacy As America'S Worst-Ever President | Dana Nuccitelli". The Guardian, 2017, https://www.theguardian.com/environment/climate-consensus-97-per-cent/2017/jun/01/donald-trump-just-cemented-his-legacy-as-americas-worst-ever-president.

30. Washington Post, 2017, https://www.washingtonpost.com/news/arts-and-entertainment/wp/2017/06/02/arnold-schwarzenegger-slams-donald-trump-over-paris-accord-decision/?utm_term=.f9374768c9ba.

31. "Rock 104.5 Talks With Tom Morello About New Music And Climate Change". KDOT, 2017, https://www.kdot.com/2017/09/15/rock-104-5-talks-with-tom-morello-about-new-music-and-climate-change/.

32. "Apple, Microsoft And Others Urge Trump To Keep Paris Climate Pact". Engadget, 2017, https://www.engadget.com/2017/06/01/apple-microsoft-google-facebook-paris-agreement-letter/.

33. Sampathkumar, Mythili. "World's Biggest Oil Companies Urge Donald Trump To Stay In Paris Climate Change Agreement". The Independent, 2017, https://www.independent.co.uk/news/world/americas/us-politics/trump-paris-agreement-climate-change-oil-companies-shell-bp-exxon-urge-president-stay-in-a7745666.html. Accessed 8 Oct 2018.

34. Abbasi, Waseem. Usatoday.Com, 2017, https://www.usatoday.com/story/news/2017/06/07/even-north-korea-slams-us-for-withdrawing-from-paris-climate-pact/102594044/.

## Chapter 4: Mechanisms, Counting Down The Days & Our Economy

1. "Global Sites Go Dark For Earth Hour". BBC News, 2015, https://www.bbc.com/news/av/world-

32103979/global-landmarks-switch-off-the-lights-for-earth-hour. Accessed 9 July 2018.

2. "Key World Energy Statistics". 2017, https://www.iea.org/publications/freepublications/publication/KeyWorld2017.pdf. Accessed 9 July 2018.

3. Harari, Yuval N. Sapiens.

4. "The World'S Richest People Emit The Most Carbon - Our World". Ourworld.Unu.Edu, https://ourworld.unu.edu/en/the-worlds-richest-people-also-emit-the-most-carbon. Accessed 9 July 2018.

5. "Norway - Energy System Overview". 2018, https://www.iea.org/media/countries/Norway.pdf. Accessed 9 July 2018.

6. "Exports Of Norwegian Oil And Gas - Norwegianpetroleum.No". Norwegianpetroleum.No, https://www.norskpetroleum.no/en/production-and-exports/exports-of-oil-and-gas/. Accessed 9 July 2018.

7. Boden, T.A., Marland, G., and Andres, R.J. (2017). National $CO_2$ Emissions from Fossil-Fuel Burning, Cement Manufacture, and Gas Flaring: 1751-2014, Carbon Dioxide Information Analysis Center, Oak Ridge National Laboratory, U.S. Department of Energy, doi 10.3334/CDIAC/00001_V2017.

8. "$CO_2$ And Other Greenhouse Gas Emissions". Our World In Data, https://ourworldindata.org/co2-and-other-greenhouse-gas-emissions#the-long-run-history-cumulative-co2. Accessed 9 July 2018.

9. "EDGAR - GHG ($CO_2$, $CH_4$, $N_2O$, F-Gases) Emission Time Series 1990-2012 Per Region/Country - European Commission". Edgar.Jrc.Ec.Europa.Eu,

http://edgar.jrc.ec.europa.eu/overview.php? v=CO2ts1990-2015. Accessed 9 July 2018.

10. "Sources Of Greenhouse Gas Emissions | US EPA". US EPA, 2018, https://www.epa.gov/ghgemissions/sources-greenhouse-gas-emissions#electricity. Accessed 9 July 2018.

11. "Sources Of Greenhouse Gas Emissions | US EPA". US EPA, https://www.epa.gov/ghgemissions/sources-greenhouse-gas-emissions#industry. Accessed 9 July 2018.

12. "Sources Of Greenhouse Gas Emissions | US EPA". US EPA, https://www.epa.gov/ghgemissions/sources-greenhouse-gas-emissions#transportation. Accessed 9 July 2018.

13. "Sources Of Greenhouse Gas Emissions | US EPA". US EPA, https://www.epa.gov/ghgemissions/sources-greenhouse-gas-emissions#commercial-and-residential. Accessed 9 July 2018.

14. Ontl, T. and Schulte, L. (2012). Soil Carbon Storage. [online] Nature.com. Available at: https://www.nature.com/scitable/knowledge/library/soil-carbon-storage-84223790 [Accessed 9 Jul. 2018].

15. Ipcc.ch. (n.d.). Emissions Scenarios. [online] Available at: http://www.ipcc.ch/ipccreports/sres/emission/index.php?idp=77 [Accessed 9 Jul. 2018].

16. Media, A. (n.d.). Deforestation and Climate Change - Climate and Weather. [online] Climateandweather.net. Available at:

https://www.climateandweather.net/global-warming/deforestation.html [Accessed 9 Jul. 2018].

17. Earthobservatory.nasa.gov. (n.d.). Tropical Deforestation. [online] Available at: https://earthobservatory.nasa.gov/Features/Deforestatio n [Accessed 9 Jul. 2018].

18. Kaye, L. (n.d.). Companies, Countries Most Responsible for Deforestation. [online] Triple Pundit: People, Planet, Profit. Available at: https://www.triplepundit.com/2015/02/breaking-ngo-lists-companies-countries-responsible-deforestation/ [Accessed 9 Jul. 2018].

19. MCSWEENEY, R. (2015). Amazon rainforest is taking up a third less carbon than a decade ago | Carbon Brief. [online] Carbon Brief. Available at: https://www.carbonbrief.org/amazon-rainforest-is-taking-up-a-third-less-carbon-than-a-decade-ago [Accessed 9 Jul. 2018].

20. US EPA. (n.d.). Sources of Greenhouse Gas Emissions | US EPA. [online] Available at: https://www.epa.gov/ghgemissions/sources-greenhouse-gas-emissions#agriculture

21. Pearce, Fred. "What Is The Carbon Limit? That Depends Who You Ask". Yale E360, 2014, https://e360.yale.edu/features/what_is_the_carbon_limi t_that_depends_who_you_ask.

22. IPCC, 2013: Summary for Policymakers. In: Climate Change 2013: The Physical Science Basis. Contribution of Working Group I to the Fifth Assessment Report of the Intergovernmental Panel on Climate Change [Stocker, T.F., D. Qin, G.-K. Plattner, M. Tignor, S. K. Allen,

J. Boschung, A. Nauels, Y. Xia, V. Bex and P.M. Midgley (eds.)]. Cambridge University Press, Cambridge, United Kingdom and New York, NY, USA.

23. Willis, Rebecca. "Paris 2015: Getting A Global Agreement On Climate Change". Green-Alliance.Org.Uk, 2014, http://www.green-alliance.org.uk/resources/Paris%202015-getting%20a%20global%20agreement%20on%20climate%20change.pdf.

24. Raupach, Michael R. et al. "Sharing A Quota On Cumulative Carbon Emissions". Nature Climate Change, vol 4, no. 10, 2014, pp. 873-879. Springer Nature, doi:10.1038/nclimate2384.

25. Hansen, James et al. "Assessing "Dangerous Climate Change": Required Reduction Of Carbon Emissions To Protect Young People, Future Generations And Nature". Plos ONE, vol 8, no. 12, 2013, p. e81648. Public Library Of Science (Plos), doi:10.1371/journal.pone.0081648.

26. McGlade, Christophe, and Paul Ekins. "The Geographical Distribution Of Fossil Fuels Unused When Limiting Global Warming To 2 °C". Nature, vol 517, no. 7533, 2015, pp. 187-190. Springer Nature, doi:10.1038/nature14016.

27. "The End Of The Oil Age". The Economist, 2003, https://www.economist.com/leaders/2003/10/23/the-end-of-the-oil-age.

28. Hutt, Rosamond. "Which Economies Are Most Heavily Reliant On Oil?". World Economic Forum, 2016, https://www.weforum.org/agenda/2016/05/which-economies-are-most-reliant-on-oil/.

29. Schipp, Debbie. "Venezuelans Forced To Scavenge In The Mud For Metal To Sell To Feed Their Families". News.Com.Au, 2018, https://www.news.com.au/finance/economy/world-economy/every-time-you-buy-less-venezuela-teeters-on-the-brink/news-story/7b651bdb5481d0450786e443d789336f.

30. "How Chávez And Maduro Have Impoverished Venezuela". The Economist, https://www.economist.com/finance-and-economics/2017/04/06/how-chavez-and-maduro-have-impoverished-venezuela.

## Chapter 5: Walking Forwards & Backwards

1. "Subsidy". Investopedia, https://www.investopedia.com/terms/s/subsidy.asp.

2. "Negative Externalities". Economicsonline.Co.Uk, 2018, http://www.economicsonline.co.uk/Market_failures/Externalities.html.

3. "Pigouvian Taxes". The Economist, 2017, https://www.economist.com/economics-brief/2017/08/19/pigouvian-taxes.

4. Coady, David et al. "How Large Are Global Energy Subsidies?". International Monetary Fund, 2015, https://www.imf.org/external/pubs/ft/wp/2015/wp15105.pdf.

5. "Lazard's Levelized Cost Of Energy Analysis — Version 11.0". Lazard, 2017,

https://www.lazard.com/media/450337/lazard-levelized-cost-of-energy-version-110.pdf.

6. Porter, Eduardo. "Does A Carbon Tax Work? Ask British Columbia". Nytimes.Com, 2016, https://www.nytimes.com/2016/03/02/business/does-a-carbon-tax-work-ask-british-columbia.html.

7. Romm, Joseph. Climate Change. Oxford University Press, 2016.

8. Meakin, Stephanie. "The Rio Earth Summit: Summary Of The United Nations Conference On Environment And Development (BP-317E)". Publications.Gc.Ca, 1992, http://publications.gc.ca/Collection-R/LoPBdP/BP/bp317-e.htm.

9. "UNFCCC". Neoias.Com, https://www.neoias.com/index.php/neoias-current-affairs/360-unfccc.

10. "Full Text Of The Convention 4". UNFCCC, 2018, http://unfccc.int/cop4/conv/conv_004.htm.

11. Seo, S. Niggol. "Beyond The Paris Agreement: Climate Change Policy Negotiations And Future Directions". Regional Science Policy & Practice, vol 9, no. 2, 2017, pp. 121-140. Wiley, doi:10.1111/rsp3.12090.

12. Cooke, Shamus. Why Copenhagen Failed. 2009, https:// https://www.globalresearch.ca/why-copenhagen-failed/16612. Accessed 9 Oct 2018.

13. "IEMA Welcomes Ambitious And Necessary Climate Change Agreement". IEMA, 2015, https://www.iema.net/news/2016/01/08/IEMA-Welcomes-Ambitious-and-Necessary-Climate-Change-Agreement-/.

14. "Emissions Gap Report 2017: Governments, Non-

State Actors Must Do More To Reach Paris Agreement". UN Environment, 2017, https://www.unenvironment.org/news-and-stories/press-release/emissions-gap-report-2017-governments-non-state-actors-must-do-more.

15. Milman, Oliver. "Paris Deal: A Year After Trump Announced US Exit, A Coalition Fights To Fill The Gap". The Guardian, 2018, https://www.theguardian.com/us-news/2018/may/31/paris-climate-deal-trump-exit-resistance.

16. "The Consequences Of Leaving The Paris Agreement". Council On Foreign Relations, 2017, https://www.cfr.org/backgrounder/consequences-leaving-paris-agreement.

17. "Countries Made Only Modest Climate-Change Promises In Paris. They're Falling Short Anyway.". Washington Post, https://www.washingtonpost.com/national/health-science/its-not-fast-enough-its-not-big-enough-theres-not-enough-action/2018/02/19/5cf0a7d4-015a-11e8-9d31-d72cf78dbeee_story.html?noredirect=on.

18. Tabuchi, Hiroko. "U.N. Climate Projects, Aimed At The Poorest, Raise Red Flags". Nytimes.Com, 2017, https://www.nytimes.com/2017/11/16/climate/green-climate-fund.html.

## Chapter 6: Impact

1. "Edward Jenner And The Development Of The First Modern Vaccine". *VBI Vaccines Inc.*, 2015,

https://www.vbivaccines.com/wire/edward-jenner-and-the-first-modern-vaccine/.

2. "Smallpox". *Our World In Data*, https://ourworldindata.org/smallpox.

3. Inglis-Arkell, Esther. *Io9.Gizmodo.Com*, 2013, https://io9.gizmodo.com/the-horrifying-story-of-the-last-death-by-smallpox-1161664590.

4. Romm, Joseph. Climate Change. Oxford University Press, 2016.

5. Johnstone, Micael. "International Agreement On HFCS A Successor To The Montreal Protocol | Lexology". *Lexology*, 2017, https://www.lexology.com/library/detail.aspx?g=1b881aa4-2de9-4c7e-968e-0aa08f3a0a07.

6. "Brazil's Success In Reducing Deforestation". *Union Of Concerned Scientists*, 2011, https://www.ucsusa.org/global-warming/solutions/stop-deforestation/brazils-reduction-deforestation.html#.W6JKpZMzYdo.

7. "Brazil 'Invites Deforestation' With Overhaul Of Environmental Laws". *The Guardian*, 2018, https://www.theguardian.com/world/2018/mar/01/brazil-amazon-protection-laws-invite-deforestation-ngo.

8. Butler, Rhett. "Deforestation Continues Upward Trend In Brazil, Says NGO". *Mongabay Environmental News*, 2018, https://news.mongabay.com/2018/08/deforestation-continues-upward-trend-in-brazil-says-ngo/.

9. "Global Grain Demand To Increase 70 Percent By 2050". *Western Farm Press*, 2011, https://www.westernfarmpress.com/markets/global-

grain-demand-increase-70-percent-2050.

10. "Options To Reduce Nitrous Oxide Emissions (Final Report)". *Ec.Europa.Eu*, 1998, http://ec.europa.eu/environment/enveco/climate_chan ge/pdf/nitrous_oxide_emissions.pdf.

11. Hermanson, Ronald et al. *Fortress.Wa.Gov*, 2018, https://fortress.wa.gov/ecy/publications/documents/00 10015.pdf.

12. "MANAGEMENT OF NITROGEN FERTILIZER TO REDUCE NITROUS OXIDE (N2O) EMISSIONS FROM FIELD CROPS/Climate Change And Agriculture Fact Sheet Series—MSU Extension Bulletin E3152 November 2014". *Michigan State University*, 2018, http://www.canr.msu.edu/uploads/resources/pdfs/ma nagement_of_nitrogen_fertiler_(e3152).pdf.

13. International, Regeneration. "What Is No-Till Farming?". *Regeneration International*, 2018, http://www.regenerationinternational.org/2018/06/25/ what-is-no-till-farming/.

14. Folnovic, Tanja. "No-Till Farming; One Step Closer To Sustainability". *Agrivi*, 2017, http://blog.agrivi.com/post/no-till-farming-one-step-closer-to-sustainability.

15. "Grass Vs. Grain". *Brittain Farm*, https://brittainfarm.com/grass-vs-grain.

16. "AMP Grazing". *Standard Soil*, http://standardsoil.com/our-approach/amp-grazing/.

17. Stanley, Paige L. et al. "Impacts Of Soil Carbon Sequestration On Life Cycle Greenhouse Gas Emissions In Midwestern USA Beef Finishing Systems". *Agricultural Systems*, vol 162, 2018, pp. 249-258. *Elsevier BV*,

doi:10.1016/j.agsy.2018.02.003.

18. "Modern Cooking Solutions: Status And Challenges". 2011, http://siteresources.worldbank.org/EASTASIAPACIFIC EXT/Resources/226300-1318878759408/eap_energy_flagship_2011_chapter4.pdf.

19. MacGuill, Dan. "How Cooking With Coal And Wood Fires Is Killing Millions Around The World". *Thejournal.Ie*, 2014, http://www.thejournal.ie/health-risks-fossil-fuels-pollution-1650076-Sep2014/.

20. "Short Report: The Benefits Of Using Compost For Mitigating Climate Change". 2011, https://www.epa.nsw.gov.au/-/media/epa/corporate-site/resources/waste/110171-compost-climate-change.pdf?la=en&hash=7ADC0B32600A8EE49E72187E4A027FA1C809AEAE.

21. Hawken, Paul. *Drawdown*. Penguin Books, 2017.

22. Jeswani, H.K., Smith, R.W. & Azapagic, A. Int J Life Cycle Assess (2013) 18: 218. https://doi.org/10.1007/s11367-012-0441-8

23. "Global Primary Energy Consumption". *Our World In Data*, 2016, https://ourworldindata.org/grapher/global-primary-energy.

24. Seger, Brian. "Global Energy Consumption: The Numbers For Now And In The Future". *DTU Orbit*, 2016, http://orbit.dtu.dk/files/128048208/Global_Energy_Consumption_The_Numbers_for_Now_and_in_the_Future.pdf.

25. Bakke, Gretchen Anna. *The Grid*. 2016.

26. "Biomass - Energy Explained, Your Guide To Understanding Energy - Energy Information Administration". *Eia.Gov*, https://www.eia.gov/energyexplained/?page=biomass_home.

27. "How Biopower Works". *Union Of Concerned Scientists*, https://www.ucsusa.org/clean_energy/our-energy-choices/renewable-energy/how-biomass-energy-works.html#.W8RBv9dvaUn.

28. "WAVE ENERGY TECHNOLOGY BRIEF". *International Renewable Energy Agency*, 2014, http://www.irena.org/documentdownloads/publications/wave-energy_v4_web.pdf.

29. "TIDAL ENERGY TECHNOLOGY BRIEF". *International Renewable Energy Agency*, 2014, http://www.irena.org/documentdownloads/publications/tidal_energy_v4_web.pdf.

30. IRENA (2017), Geothermal Power: Technology Brief, International Renewable Energy Agency, Abu Dhabi.

31. Ahearn, Sean. "El Salvador: The "Land Of Volcanoes" And Geothermal Energy". *Worldwatch Institute*, http://www.worldwatch.org/el-salvador-"land-volcanoes"-and-geothermal-energy.

32. "Geothermal Energy In Philippines | REVE". *Evwind.Es*, 2012, https://www.evwind.es/2012/08/31/geothermal-energy-in-philippines-2/22730.

33. Stockton, Nick. "Nuclear Power Is Too Safe To Save The World From Climate Change". *WIRED*, 2016,

https://www.wired.com/2016/04/nuclear-power-safe-save-world-climate-change/.

34. Ritchie, Hannah. "What Was The Death Toll From Chernobyl And Fukushima?". *Our World In Data*, 2017, https://ourworldindata.org/what-was-the-death-toll-from-chernobyl-and-fukushima.

35. Johnson, Nathanael. "Meltdowns, Waste, And War: Here Are The Real Risks Of Nuclear". *Grist*, 2018, https://grist.org/article/nuclear-is-scary-lets-face-those-fears/.

36. Barzashka, Ivanka. "Converting A Civilian Enrichment Plant Into A Nuclear Weapons Material Facility". *Bulletin Of The Atomic Scientists*, 2013, https://thebulletin.org/2013/10/converting-a-civilian-enrichment-plant-into-a-nuclear-weapons-material-facility/.

37. "Hydropower Technology Brief". *International Renewable Energy Agency*, 2015, http://www.irena.org/documentdownloads/publications/irena-etsap_tech_brief_e06_hydropower.pdf.

38. "Lazard's Levelized Cost Of Energy Analysis — Version 11.0". Lazard, 2017, https://www.lazard.com/media/450337/lazard-levelized-cost-of-energy-version-110.pdf.

39. "Wind Power Technology Brief". *International Renewable Energy Agency*, 2016, http://www.irena.org/-/media/Files/IRENA/Agency/Publication/2016/IRENA-ETSAP_Tech_Brief_Wind_Power_E07.pdf.

40. "Wind In Power 2010 European Statistics". *EWEA*, 2011,

http://www.ewea.org/fileadmin/ewea_documents/documents/statistics/EWEA_Annual_Statistics_2010.pdf.

41. "Annual Electricity Generation In Germany". *Fraunhofer ISE*, 2018, https://www.energy-charts.de/energy.htm?source=all-sources&period=annual&year=all.

42. "The Wind Industry Is A Strong Employer In Germany". *VDMA*, https://www.vdma.org/documents/106078/16262656/1491380194390_VDMA%20PS%20BWE%20OWIE%20Ent-PM-GWS%20BeschAnaly%202017-03-22-final%20E.pdf/73dc4b40-2fd4-4971-a615-51472ea135b8. Accessed 9 Aug 2018.

43. Buchsbaum, L. "Mixed Mandate: Germany'S New Coal Commission Struggles To Balance Environment And Jobs". *Energy Transition*, 2018, https://energytransition.org/2018/06/mixed-mandate-germanys-new-coal-commission-struggles-to-balance-environment-and-jobs/.

44. "How Much Power Does The Sun Give Us? | Solar Powered In Toronto". *Yourturn.Ca*, http://www.yourturn.ca/solar/solar-power/how-much-power-does-the-sun-give-us/.

45. Chandler, David L. "Vast Amounts Of Solar Energy Radiate To The Earth, But Tapping It Cost-Effectively Remains A Challenge". *Phys.Org*, 2011, https://phys.org/news/2011-10-vast-amounts-solar-energy-earth.html.

46. "The History Of Solar Power". *Experience*, 2017, https://www.experience.com/advice/careers/ideas/the-history-of-solar-power/.

47. "The History Of Solar". *US Department Of Energy,* https://www1.eere.energy.gov/solar/pdfs/solar_timelin e.pdf.

48. Guseff, Alex. "SOLAR POWER". *Techno Tomorrow,* http://technotomorrow.com/en/energy/solar-power.html.

49. Shahan, Zachary. "Solar Panel Cost Trends (Tons Of Charts)".          *Cleantechnica,*          2014, https://cleantechnica.com/2014/09/04/solar-panel-cost-trends-10-charts/.

50. O'Boyle, Michael, and Silvio Marcacci. "Wind And Solar Costs Continue To Drop Below Fossil Fuels. What Barriers Remain For A Low-Carbon Grid?". *Utility Dive,* 2018,    https://www.utilitydive.com/news/wind-and-solar-costs-continue-to-drop-below-fossil-fuels-what-barriers-rem/519671/.

51. Harrington, Rebecca, and Tech Insider. "Here's How Much Of The World Would Need To Be Covered In Solar Panels To Power Earth". *Business Insider,* 2015, https://www.businessinsider.com/map-shows-solar-panels-to-power-the-earth-2015-9.

52. European Commission, Joint Research Centre (JRC). "Urbanization: 95% Of The World's Population Lives On 10% Of The Land." ScienceDaily. ScienceDaily, 19 December 2008. <www.sciencedaily.com/releases/2008/12/081217192745.htm>.

53. "Living In Peace On Our Planet: Insight, Energy, Environment".          *Planetthoughts.Org,* http://www.planetthoughts.org/? pg=vid/ShowVideo&qid=3149.

54. Smith, Andrew ZP. "Fact Checking Elon Musk'S

Blue Square: How Much Solar To Power The US? - UCL Energy Institute Blog". *University College London Energy Institute,* 2015, http://blogs.ucl.ac.uk/energy/2015/05/21/fact-checking-elon-musks-blue-square-how-much-solar-to-power-the-us/.

55. "The Footprint Of Coal". *Sourcewatch,* https://www.sourcewatch.org/index.php/The_footprint_of_coal.

56. Allred, B. W. et al. "Ecosystem Services Lost To Oil And Gas In North America". *Science,* vol 348, no. 6233, 2015, pp. 401-402. *American Association For The Advancement Of Science (AAAS),* doi:10.1126/science.aaa4785.

57. STEFFY, LOREN. "The Generation Gap". *Texas Monthly,* 2014, https://www.texasmonthly.com/articles/the-generation-gap/.

58. Rehman, Shafiqur et al. "Pumped Hydro Energy Storage System: A Technological Review". *Renewable And Sustainable Energy Reviews,* vol 44, 2015, pp. 586-598. *Elsevier BV,* doi:10.1016/j.rser.2014.12.040.

59. "Hydrogen: A Promising Fuel And Energy Storage Solution". *National Renewable Energy Laboratory (NREL) Continuum,* https://www.nrel.gov/continuum/energy_integration/hydrogen.html.

60. "Compressed Air Energy Storage (CAES)". *Energy Storage Association,* http://energystorage.org/compressed-air-energy-storage-caes.

61. Mallela, Venkateswara Sarma et al. "Trends In

Cardiac Pacemaker Batteries". *The Indian Pacing And Electrophysiology Journal,* 2004, https://www.researchgate.net/publication/6844770_Trends_in_Cardiac_Pacemaker_Batteries.

62. Hart, David M. et al. "Energy Storage For The Grid: Policy Options For Sustaining Innovation". *MIT Energy Initiative (MITEI),* 2018, https://energy.mit.edu/wp-content/uploads/2018/04/Energy-Storage-for-the-Grid.pdf.

63. "Lithium-Ion Batteries For Large-Scale Grid Energy Storage (ESS)". *Research Interfaces,* 2018, https://researchinterfaces.com/lithium-ion-batteries-grid-energy-storage/.

64. Clover, Ian. "Lithium-Ion Batteries Below $200/Kwh By 2019 Will Drive Rapid Storage Uptake, Finds IHS Markit". *PV Magazine International,* 2017, https://www.pv-magazine.com/2017/08/03/lithium-ion-batteries-below-200kwh-by-2019-will-drive-rapid-storage-uptake-finds-ihs-markit/.

65. "The History Of The Electric Car". *Department Of Energy,* 2014, https://www.energy.gov/articles/history-electric-car.

66. Adams, Eric. "The Age Of Electric Aviation Is Just 30 Years Away". *WIRED,* 2017, https://www.wired.com/2017/05/electric-airplanes-2/.

67. Hanley, Steve. "China Launches World's First All-Electric Cargo Ship, Will Use It To Haul Coal". *Cleantechnica,* 2017, https://cleantechnica.com/2017/12/02/china-launches-worlds-first-electric-cargo-ship-will-use-haul-coal/.

68. Boffey, Daniel. "World's First Electric Container

Barges To Sail From European Ports This Summer". *The Guardian,* 2018, https://www.theguardian.com/environment/2018/jan/24/worlds-first-electric-container-barges-to-sail-from-european-ports-this-summer.

69. "Hydrogen Energy". *Renewable Energy World,* https://www.renewableenergyworld.com/hydrogen/tech.html.

70. CORNEIL, HAMPTON G., and FRED J. HEINZEL-MANN. "Hydrogen In Oil Refinery Operations". *ACS Symposium Series,* 1980, pp. 67-94. *AMERICAN CHEMICAL SOCIETY,* doi:10.1021/bk-1980-0116.ch005.

71. Heidrich, Elizabeth S. et al. "Performance Of A Pilot Scale Microbial Electrolysis Cell Fed On Domestic Wastewater At Ambient Temperatures For A 12 Month Period". *Bioresource Technology,* vol 173, 2014, pp. 87-95. *Elsevier BV,* doi:10.1016/j.biortech.2014.09.083.

72. CORNEIL, HAMPTON G., and FRED J. HEINZEL-MANN. "Hydrogen In Oil Refinery Operations". *ACS Symposium Series,* 1980, pp. 67-94. *AMERICAN CHEMICAL SOCIETY,* doi:10.1021/bk-1980-0116.ch005.

73. "HY4: Maiden Flight Of The Hydrogen Powered Airplane". *Aviation Pros,* 2016, https://www.aviationpros.com/press_release/12264576/hy4-maiden-flight-of-the-hydrogen-powered-airplane.

74. Brian, Matt. "Easyjet Is Serious About Electric Planes". *Engadget,* 2017, https://www.engadget.com/2017/10/03/easyjet-innovation-day-tech-electric-planes/.

75. "Easyjet Showcases Airbus A320neo At Manchester Airport". *Breaking Travel News,* 2018,

http://www.breakingtravelnews.com/news/article/easy jet-showcases-airbus-a320neo-at-manchester-airport/.

76. Topham, Gwyn. "Rolls-Royce Joins Race To Develop Electric Passenger Jets". *Guardian*, 2017, https://www.theguardian.com/business/2017/nov/28/r olls-royce-electric-passenger-jets-airbus-siemens-e-fan-x-hybrid-electric-plane-2020.

77. Bennett, Jay. "NASA's Next Great X-Plane Will Try To Revolutionize Electric Flight". *Popular Mechanics*, 2017, https://www.popularmechanics.com/flight/a26609/nas a-plane-electric-x-57-maxwell/.

78. "A Partial List Of Products Made From Petroleum". *Whgbetc.Com*, https://whgbetc.com/petro-products.pdf.

79. "Here's What Oil At $70 Means For The World Economy". *Bloomberg*, 2018, https://www.bloomberg.com/news/articles/2018-05-09/here-s-what-oil-at-70-means-for-the-world-economy.

80. Schwartz, Tyler. "US Paraxylene Prices Top $1,300/Mt For First Time Since 2014 | S&P Global Platts". *S&P Global*, 2018, https://www.spglobal.com/platts/en/market-insights/latest-news/petrochemicals/083018-us-paraxylene-prices-top-1300mt-for-first-time-since-2014.

81. Keith, David W. et al. "A Process For Capturing $CO_2$ From The Atmosphere". *Joule*, vol 2, no. 8, 2018, pp. 1573-1594. *Elsevier BV*, doi:10.1016/j.joule.2018.05.006.

82. Fuss, Sabine et al. "Betting On Negative Emissions". *Nature Climate Change*, vol 4, no. 10, 2014, pp. 850-853. *Springer Nature*, doi:10.1038/nclimate2392.

83. "GLOBAL ENERGY TRANSFORMATION: A

ROADMAP TO 2050". *International Renewable Energy Agency*, 2018, http://www.irena.org/-/media/Files/IRENA/Agency/Publication/2018/Apr/I RENA_Report_GET_2018.pdf.

84. Roberts, David. "Adapting To Climate Change: Necessary But Difficult And Expensive". *Grist*, 2012, https://grist.org/climate-change/adapting-to-climate-change-necessary-but-difficult-and-expensive/.

## Conclusion

1. "About 350.Org". *350.Org*, https://350.org/about/.
2. "About Us | Beyond Coal". *Sierra Club*, 2018, https://content.sierraclub.org/coal/about-the-campaign.
3. "About CAN | CAN International". *Climatenetwork.Org*,
http://www.climatenetwork.org/about/about-can.

# ABOUT THE AUTHOR

Mohammed Ghassan Farija is a Bahraini writer, environmental and mechanical engineer, and the author of *The Layman's Guide To Climate Change*. He holds a Master's degree in Environmental Engineering from Newcastle University and Bachelor's degree in Mechanical Engineering from Manchester Metropolitan University. Since completing his  Master's in 2015, he has gone on to work as a freelance environmental consultant, a project engineer, and currently as an environmental engineer. His writing covers a broad range of topics, ranging from science to philosophy to religion.

Made in the USA
San Bernardino, CA
20 February 2019